Burt Green Wilder

Physiology Practicums

Burt Green Wilder

Physiology Practicums

ISBN/EAN: 9783744725781

Printed in Europe, USA, Canada, Australia, Japan

Cover: Foto ©berggeist007 / pixelio.de

More available books at **www.hansebooks.com**

"I have been in the habit of advising my students to dissect the CAT as a convenient preliminary to practical Human Anatomy."—*Joseph Leidy.*

"It seems to me that the first dissections should be made on CATS and dogs until a good techni been acquired, so that the supply of human cadavers, which is always insufficient, can be fully utilized to the best advantage."—*J. S. Billings.*

"There is so close a solidarity between ourselves and the animal world that our inaccessible inward parts may be supplemented by theirs. A SHEEP'S heart or lungs or eye must not be confounded with those of man ; but so far as the comprehension of the elementary facts of the physiology of circulation and of respiration and of vision goes, the one furnishes the needed anatomical data as well as the other."— *Huxley.*

PHYSIOLOGY PRACTICUMS

EXPLICIT DIRECTIONS FOR EXAMINING

PORTIONS OF THE CAT, AND THE HEART, EYE,

AND BRAIN OF THE SHEEP

AS AN AID IN THE

STUDY OF ELEMENTARY PHYSIOLOGY

SECOND EDITION, REVISED

WITH THIRTY FIGURES

BY

BURT G. WILDER, B.S., M.D.,

Professor of Physiology, Vertebrate Zoölogy, and Neurology in Cornell University ; formerly Professor of Physiology in the Medical School of Maine and the University of Michigan ; President (1885) of the American Neurological Association and of the Biological Section of the Amer. Association Adv. Science, etc.

PUBLISHED BY THE AUTHOR

PRESSES OF THE ITHACA JOURNAL.

1895

PREFACE TO THE FIRST EDITION.

About ten years ago, in the effort to enable the members of the general class in Physiology at Cornell University (150–180 in number) to study for themselves intelligently certain parts of the cat and sheep as an aid to the comprehension of the functions and relations of the corresponding human organs, I put alcoholic specimens before them and wrote on the blackboard brief directions which were orally amplified and illustrated. A few years later these directions were written upon cloth sheets that were suspended before the class. They were amplified and printed in the fall of 1889 and issued in their present form in 1892.

The separation of the sheets and plates has obvious inconveniences but upon the whole the practical advantages are greater.

From the first the assistants and students have cordially coöperated toward increasing accuracy and explicitness.

It is to be hoped that ere long as much as is here included may be required for admission to this and other universities, so that the instruction therein may commence upon a foundation both higher and more substantial than at present.

Ithaca, N. Y., December 26, 1893.

PREFACE TO THE SECOND EDITION.

The text has been revised and largely rewritten. An effort has been made to correct the errors and omissions detected during the three years' use of the work at Cornell University and elsewhere. For helpful suggestions I am particularly indebted to my assistants, Dr. P. A. Fish and Dr. B. B. Stroud.

The changes in the illustrations comprise new figures of the cat's skeleton, and of the sheep's heart and brain. Two outlines have been introduced into the text.

The order has been modified so as to bring the examination of the head and neck just before that of the eye and brain. The eleven practicums are combined so as to form four Parts, each dealing with a natural group of subjects.

A teaching experience of twenty-seven years leads me to believe that explicitness should be a main feature of directions for beginners. To credit them with unlikely knowledge, judgment and skill, or with inspiration that will serve in place of those attributes, may compliment them and simplify the task of the writer. But there result perplexities, the formation of faulty methods, and the waste of time and material.

When, however, there has once been established a sound basis of fact and manipulation, the student may safely and profitably venture upon unfamiliar ground. He may either apply the directions to different forms, or re-examine the same forms in different ways. For example, the brain of the cat, dog, monkey or man may compared with that of the sheep, and the sheep's brain may be explored in ways other than that presented in the following pages.

September 20, 1895.

CONTENTS.

LIST OF PLATES.

FIGURES IN THE TEXT.

PHYSIOLOGY PRACTICUMS.

PART I.

PRACTICUM I: THE CAT: ITS FORM AND CERTAIN PARTS OF ITS STRUCTURE.

PLATES REQUIRED: I—IV.

§ 1. *Comparison of the Cat with Man.*—At one of the earlier lectures of the course the cat's form, attitude and mode of progression, and the main features of its skeleton, were compared with those of man. Examine the mounted skeleton. Manipulate the specimen. Press upon the regions where bony prominences exist. Move the limbs as wholes and at their joints. Verify the statements made at the lecture and note additional points of resemblance and difference if possible.

a. The preserved specimen is less well-adapted for these topographic observations than the freshly-killed animal. Still more may be learned from the living cat, provided it and the observer are on such confidential terms as to permit unrestricted manipulation.

§ 2. *The Leg.*—Recognize the three JOINTS, proximal, the HIP, distal, the ANKLE, and intermediate, the KNEE, demarcating as many segments, *viz*, the THIGH with its single bone, FEMUR ; the LEG proper (sometimes called *crus*) with its two bones TIBIA and FIBULA, and the FOOT (pes) composed of several small bones.

§ 3. One or both of the heels should retain a piece of the *tendo Achillis*, seen on the right in Fig. 1. Also at the knee should be retained the PATELLA or "knee-pan" (Pl. I and Fig. 1) a movable bone attached by a strong ligament to the head of the tibia, and giving insertion to the muscles on the "front" of the thigh.

§ 4. At the sides of the left patella cut carefully into the knee joint. Then cut transversely so as to separate the leg proper from the thigh. Note that the apposed ends of the femur and tibia present a bluish white covering of CARTILAGE (gristle). This forms an elastic cushion like a buffer, to lessen the shock in moving and especially in alighting from a height.

a. Shave off a thin slice of cartilage ; hold it to the light and note its translucence.

b. In a freshly killed animal the interior of the joints would present a moist surface due to the SYNOVIA secreted by the lining membrane.

§ 5. *Removing the Thigh.*—*a.* Move the left thigh to and fro so as to indicate where its bone, the FEMUR, joins the pelvis. On Pl. I observe the irregular form of the pelvis. Use the *scissors* to carefully cut away the muscles that have been left attached to the femur and pelvis.

a. After the flesh about the proximal end of the femur is mostly removed, moving it will show that it has a HEAD imbedded in a socket, and that there is a fibrous CAPSULE extending from the margin of the socket over part of the head.

b. Push the femur as far as it will go in any direction ; this will render the capsule tense at the opposite side ; cut it here and continue till the head may be extracted slightly from the socket.

c. Note that its complete removal is prevented by a fibrous cord connecting it with the bottom of the socket ; this is the ROUND LIGAMENT of the hip joint, present also in man and many animals but absent in the orang which uses its short leg freely as an arm. Cut the ligament.

§ 6.   Compare the two ends of the femur.   The subspherical head, on the constricted neck, enters into the composition of a "ball-and-socket joint ;" the distal end forms at the knee a "hinge-joint."   The patella (knee-pan) has been removed.   The movements at either end are similar to those in man.

§ 7.   *The Periosteum.*—Near the distal end of the bone note the covering of fibrous membrane, the PERIOSTEUM.   Divide it at any point, preferably with a pocket-knife or arthrotome.   Insert the tracer between it and the bone ; strip it from the bone for a considerable area.   Near the ends of the bone there may be seen vessels passing from it into the bone ; in the dried bone the small orifices for these vessels may be detected.

§ 8.   *The Marrow.*—Transect the femur with saw, nippers or hatchet. The shaft of the bone forms a tube whose cavity is filled with a kind of fat, the MARROW.   The ends are solid but of a spongy texture.

§ 9.   Remove the right leg at the hip.   Trim off remnants of flesh with the scissors and preserve for use at Pr. II.

§ 10.   *The Skin Muscles.*—On Pl. II, the irregular lines crossing between the words THORAX and ABDOMEN indicate the cut edge of a thin muscle the caudal part of which is supposed to have been removed with the skin ; the cephalic part narrows to be attached to certain muscles of the arm.

a. In the cat, as in most quadrupeds, in addition to the ordinary muscles of the limbs and those of the trunk which are attached to bones, there is a nearly continuous sheet of muscle in close relation with the skin ; this enables the horse, for instance, to shake off a fly, while the attachment to the arm increases the efficiency of that limb in ordinary locomotion or climbing.   In man the skin muscles are present only on the neck and head, mainly as organs of expression.

§ 11.   At about the middle of the left side of the thorax, as indicated on Fig. 1, make two incisions crossing one another at right angles and about 5 cm. (2 in.) long.   Together they constitute a *crucial incision.*

a. These first incisions should divide only the skin, an apparently single, tough layer.

§ 12.   There are thus established four triangular flaps of skin. With the forceps grasp the corner of one of these flaps, lift it and with the scalpel dissect it from the subjacent parts.   Even if the incisions have divided more than the skin the latter may be isolated by taking care to lift only a *single layer* of tissue, the ental surface of which is white or dark, but not red or pink.

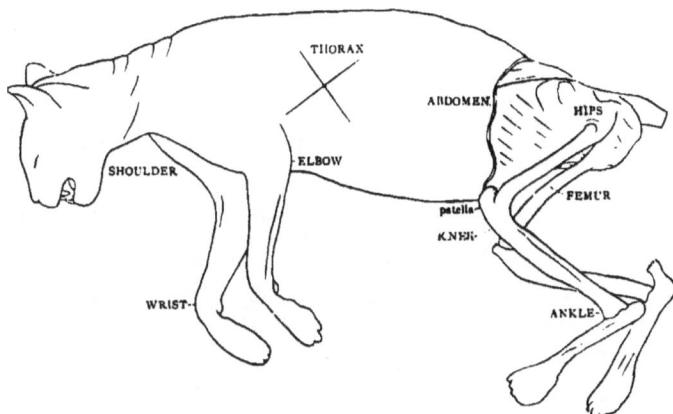

Fig. 1. Diagram of a Cat as prepared for practicum purposes. The hair was removed by immersion in water at about 80° C. (176° F.). The skin and most of the flesh have been cut from the legs and hips, and the tail abbreviated. At the right heel is seen a part of the TENDO ACHILLIS, the large tendon by which the muscles of the "calf of the leg" were inserted. The special object of the figure is to indicate the crucial incision by means of which the skin muscle is to be exposed.

a. When the four flaps of skin have been lifted they may be cut off with the scissors. There will then be exposed a quadrangular area of the SKIN-MUSCLE, recognizable by its pale red color and from the sparseness of its fascicles (bundles of fibers). Even if it has been divided by the crucial incision it may be cut out as a thin sheet, separated by FAT and CONNECTIVE TISSUE from an ental, thicker muscle, the LATISSIMUS, whose fascicles run in nearly the same direction ; see Pl. II, just dorsad of the word ABDOMEN.

b. If the foregoing operations fail on the left side, repeat on the right with additional precautions.

c. On the neck, from a point about midway of its length and dorso-ventral diameter (on Pl. II coinciding nearly with the dorsal end of the line between the E and the C of NECK), cut caudad along the middle of the left side to the root of the tail, or to the cut margin of the skin in case it and the flesh have been removed from the hips and legs. This incision should divide both the skin and the skin-muscles. On the neck and shoulder no great harm would result from cutting too deeply, but on the thorax care should be taken not to cut the latissimus, and on the abdomen there is danger of opening that cavity prematurely.

d. Grasp the cut edge of skin and skin-muscle, at the dorsal side of area exposed in (*a*), and dissect them up from the latissimus, remembering that the latter has a free margin extending obliquely toward the arm in continuation of the line shown in Pl. II. On the neck and shoulders some irregularities and adhesions will be encountered, but no serious difficulties. In this way "skin" the left side dorsad as far as the meson.

§ 13. Do the same for the ventral half of the left side, noting three features : (2) The MAMMARY GLAND, a whitish, lobulated organ, quite large in nursing females, and extending the whole length of the abdomen and upon the thorax ; (2) the series of NIPPLES connected therewith ; (3) the narrowing and thickening of the latissimus near the arm ; (4) the PECTORAL MUSCLES, thicker and darker than the skin-muscles and having a direction from the meson latero-cephalad (Pl. III).

a. Taking care not to cut the pectorals, divide the skin and skin-muscles around the arm just proximad of the elbow. Do the same for the right arm and then, if necessary, for the legs, at about the middle of the thighs. Finally cut the skin just caudad of the ears.

b. If preferred the skin may now be divided with the scissors along both mesons (dorsal and ventral median lines) ; or it may be removed from the right side by dissecting it up from either or both mesons. The skin of the head should be left until that region is dissected.

c. Examine the ental aspect of the skin and note the pale sheet of skin-muscle, with close adhesions, especially caudad ; it narrows and thickens cephalad and is connected with muscles attached to the arm.

§ 14.   Place the cat on its back and tie the arms outstretched to the loops at the side of the tray. Compare Pl. III. Recognize the convex shoulders and the axillas (arm-pits).

The MESON (middle line) may not be perfectly straight on account of the twisting of the body. At the meson may be recognized the EPI-GASTRIUM (pit of the stomach), and the STERNUM (breast-bone) extending thence to the root of the neck Pl. II, *1*.

§ 15.   *The Pectoral Muscles.*—These form, as in man, a considerable mass covering the ventral aspect of the thorax and extending thence to the brachium. Read the description on Pl. III, and note the differences from the human.

The cephalic margin of the pectoral mass overlaps the muscles on the ventral aspect of the neck ; it extends nearly transversely to the convexity of the shoulder. The cephalic margin is sometimes distinct, but sometimes quite thin and so closely adherent to the cervical muscles as to separate with some difficulty. The caudal margin is thinner and extends obliquely from the axilla to the epigastrium.

§ 16.   *Transecting the Pectorals.*—Compare Plates III and IV.

a. At a point on the cephalic margin about (not more than) one-fourth of the distance from the meson to the convexity of the shoulder, begin an incision which is to be carried caudad parallel with the meson to the caudal margin ; it should be not more than 3 mm. (one-eighth inch) deep ; the division of the ectal layers will expose ental layers not quite parallel therewith.

b. Divide the ental layers by a second incision of about the same depth, so as to reach a muscle, the RECTUS, whose direction is parallel with the meson instead of at an angle with it.

c. Just ental of the pectorals are some *nerves* and *blood-vessels* which are not to be cut at this time.

§ 17.   *The Axillary Parts.*—Lift the lateral portion of the pectoral mass and partly tear, partly cut it toward the arm. In the interval between it and the arm and the thorax may be seen :

a. Web-like CONNECTIVE TISSUE, composed of delicate fibers.

b. In well-nourished animals, masses of firm white FAT.

c. Several NERVES, white and solid.

d. A smaller number of BLOOD-VESSELS, elastic, hollow, and sometimes containing blood. The main ARTERY usually lies a little cephalad of the VEIN (Pl. 4), is more elastic and probably more nearly empty.

e. The axillary parts should be exposed with the tracer rather than the scalpel.

§ 18.   *The Clavicle* (collar-bone).—Note its location on Pl. I. Feel for it among the transected muscles at the shoulder ; it is usually laterad of the cut as in Pl. IV. To expose it more fully evert the muscles, tearing the connective tissue. When the clavicle is found tear and cut it from the muscles. Scrape clean with a dull edge. Note the slight curvature of its ends in opposite directions, and that the lateral or scapular end is the wider.

   a. In man and some other animals the clavicle is firmly connected with the sternum at one end and at the other with the scapula (shoulder-blade), and thus serves to brace the arm away from the trunk. In the cat the attachments are loose ; in the dog it is an insignificant ossicle ; in the horse, cow, sheep, etc., it is altogether absent.
   b. The cleaned and dried clavicle of the cat is entirely inoffensive and may be preserved or even carried in the vest pocket as a good example of the organs which are apparently functionless and are hence called *atelic*.

§ 19. Transect the other pectorals and axillary parts   Cut the cords that hold the arms to the tray. Turn the cat so that it rests on the ventrum. Note the wide muscle converging from the ribs to the vertebral margin of the scapula. Together these muscles support the trunk between the arms somewhat as would a broad sling.

§ 20. Divide these muscles and any other parts and detach the arms. Tie them and the right leg together with a tag bearing the name of the student, the number of his Practicum Section (I, II, III or IV) and the number of his seat (1, 2, 3 etc.).

A ligamentous skeleton of the arm should be available for examination if possible. If all are supplied contiguous students should have a right and a left. A leg was saved from Pr. I.

§ 1. *Review the Segments, Joints and Larger Bones of the Arm* from Plates I and II ; the scapula now appears to constitute a segment of the limb instead of a part of the trunk.

§ 2. *Learn the Technical Names of the Five Digits (fingers) ;* POLLEX (thumb) ; INDEX (forefinger) ; MEDIUS (middle-finger) ; ANNULARIS (ring-finger) : MINIMUS (little-finger). On Pl. V all are visible and named excepting the minimus. The pollex is the shortest, its attachment is most proximal and in the cat it is not opposable to the other digits like the human thumb.

§ 3. *Determine whether the Arm be Right or Left as follows :* Hold it in front of you, the hand downward and the elbow toward you ; if the pollex is toward your right the arm is the left ; if toward your left the arm is the right.

§ 4. *Determine the Aspects of the Arm.*—That which bears the pollex is next the thorax, and is the "inner" or ulnar ; the other is the "outer" or radial.

a. There is liability to some confusion here since the pollex is in line with the radius ; but in the cat the radius is crossed upon the ulna so as to rotate (pronate) the hand and bring the pollex upon the ulnar side.

b. The student may illustrate the two conditions upon his own arm as follows : If the hand be placed upon the table with the palm upward the pollex is "outer" and the two bones of the antibrachium (ulna and radius) are parallel, the radius as a whole lying upon the "outer" side. But if the palm be turned down the distal end of the radius crosses to the "inner" side of the ulna and the pollex likewise comes to be on the ulnar or "inner" side of the limb, as in the cat.

§ 5. *Remove the Skin from the Arm.*—In Pract. I, § 4, *h,* the skin was cut around the arm between the elbow and shoulder. It may be "stripped" to the wrist, everted as a closely-fitting glove-finger may be turned inside out. Cut it at or distad of the wrist. Its complete removal may require slitting along the dorsum of each digit.

§ 6. *With the Scissors Trim off the Muscles* remaining attached to the dorsal or vertebral border of the scapula. Since the dorsal border of the scapula is convex the concavity of the curved scissors should be applied to it. The ragged remnants of the skin-muscle attached to the latissimus should also be cut off.

§ 7. With young cats the border of the scapula consists of a strip of cartilage which later is converted into bone. With the arthrotome or a pocket-knife cut a notch through the cartilage into the bone ; the former may be broken from the latter along their line of junction.

§ 8. *Note the Remnant of the Pectoral Mass* mostly on the cephalic and "outer" aspect of the brachium ; on the caudal and "inner" side is another mass consisting of the latissimus and the TERES, a shorter muscle extending from the caudal margin of the scapula to the humerus. If the latissimus and teres be drawn entad from the brachium there will appear, passing from them to beyond the elbow a wide and strong FASCIA, almost a tendon ; between it and the brachium is connective tissue which may be cut or torn.

a. On the ental side of the shoulder, in a ragged mass between the pectoral and latissimus are a NERVE, a BLOOD-VESSEL, and connective tissue. Into the space loosely filled by them crowd the thumb or a scalpel handle. The space will be found bounded by a MUSCULO-FIBROUS ARCH.

9. *Cut the Arch at its Highest Point,* between the pectoral and latissimus masses. Turn these in opposite directions. Note that they covered a fusiform, slender muscle, the BICEPS, Pl. V, and that along its inner side, near the humerus, passes a NERVE, the musculo-cutaneous ; see § 18.

§ 10. *On the Ectal Surface of the Scapula* is a thin sheet of muscle and fascia attached to the bony ridge (spine) which projects from it. Cut along the line of this attachment and remove.

§ 11. *Removal of Other Muscles.*—Keeping the biceps constantly in mind, the muscles about the shoulder may be removed as follows :

a. Transect the mass covering the *ental aspect of the scapula* at about the middle of its length, *i. e.*, at the word *muscle* in Pl. V ; after dividing the muscle with the scalpel use the arthrotome and cut down to the bone. Peel the proximal (dorsal) part of the muscle off, noting that :

1. The fibers diverge toward the vertebral border.
2. They arise not by ordinary tendons but directly from the periosteum. The periosteum may be lifted on the point of the scalpel.

b. On the *lateral aspect* is a bony ridge, the "spine" of the scapula, shown but not named on Pl. I. Avoiding that, the muscles at either side of it are to be transected at about one-third the way from the dorsal border. Note that they arise from the "spine," as well as from the general surface of the scapula.

§ 12. *Relation of the Muscles to the Shoulder Joint.*—Cut between the distal portions of the three muscles along the margins of the scapula and peel them off to the shoulder. Their attachments cover the shoulder joint so as to lessen the liability to dislocation of the head of the humerus from the shallow glenoid cavity. Note that pulling upon them moves the arm in corresponding directions : also that various combinations of the three muscles enable the arm to be rotated in intermediate directions.

§ 13. *The Capsule of the Shoulder.*—Transect all the muscles at the head of the humerus so as to expose in some degree the fibrous capsule of the joint, comparable with what was seen at the hip ; Pract. I, § 5, *b.*

§ 14. *The Triceps.*—The "elbow" side (dorsum) of the brachium presents a thick mass of mucles which may here be designated under the general name TRICEPS.

§ 15. *The Brachial Plexus.*—On the ental side of the arm are several white, thick, firm NERVES. These are part of the Brachial Plexus, Pr. I, § 17, *c*, Pl. IV, and were cut in removing the arm.

§ 16. *The Ulnar Nerve.*—In the interval between the biceps and the triceps, on the ulnar ("inner") side, search with the tracer for a large

white cord, the ULNAR NERVE. Trace it to the elbow, where it is lodged
in a notch between the projecting olecranon process (Pl. I) and the ulnar
projection of the humerus.

    a. It has the same location in man and, notwithstanding the protec-
tion from injury thus afforded, an unguarded blow upon this part of the
elbow may so far press upon the nerve as to cause the tingling and more
or less painful sensation known as "crazy-bone."

    b. Slit the fascia of the antibrachium and separate the muscles so as
to follow the nerve distad. At about the middle of the length of the antibrachium it divides into two branches of which one is distributed to the
palm and the other to the dorsal skin of the hand ; branches are given off
to the pad at the wrist joint.

    § 17. *Division of the Triceps.*—Transect it at about the middle ; turn
the distal half away from the bone, *leaving it attached* for later observa-
tion. Cut the proximal half from the humerus and scapula.

    § 18. *The Musculo-Cutaneous Nerve.*—Between the biceps and the
bone push first a scalpel-handle and then a finger ; then draw the muscle
gently from the bone and rotate it slightly. This will expose a NERVE,
the MUSCULO-CUTANEOUS, which emerges from the parts about the shoulder, sends several branches to the proximal part of the biceps and then
continues to the elbow, passes to the radial side of the arm, and is distributed to·the skin ; *Anatomical Technology*, Fig. 103, *N. Mct.*, and § 1022.

    a. Like many other nerves it supplies both muscle and skin, con-
taining fibers derived from both the ventral (motor or "anterior") and
dorsal (sensory or "posterior") spinal nerve roots ; it is a "mixed" nerve;
see Practicum VI, the MYEL.

    § 19. *Origin of the Biceps.*—The tapering proximal end of the mus-
cle terminates in a single, subcylindrical tendon (sinew) which enters a
groove in the bone covered by a fibrous sheet ; Pl. V, *3, 4.* Slit the sheet
and follow the tendon to its attachment on a slight projection of the lip of
the glenoid cavity of the scapula.

    a. The second tendon of origin in man from the tip of the coracoid process (whence
the name *biceps*) is absent in the cat ; Pl. V.

    § 20. *The Brachialis.*—Partly covered by the biceps and lying closely
against the "outer" side of the humerus is the BRACHIALIS (*brachialis
anticus*). The distal ends of the two muscles may be followed together
as directed in § 21.

    § 21. *Insertion of the Biceps.*—This is hidden by muscles of the
antibrachium arising from both sides of the humerus but especially from
the radial ("outer") side. These must be cut away, though not neces-
sarily so far distad as in Pl. V. Transect the muscles of the antibrachium covering the insertion of the biceps at about one-third of the way
from the elbow to the wrist, and peel the proximal portions toward the
wrist, but without cutting them away at their origins.

    a. The fibrous sheath, FASCIA, covering them really receives a slip of ten-
don from the biceps. This really constitutes a second and considerable
insertion of the muscle, so that, especially in the absence of the second
head one might well regard this muscle in the cat as a *bipes* rather than a
biceps.

    § 22. *Insertion of the Brachialis.*—Use the tracer as much as possible,
the scalpel sparingly, and ascertain that the brachialis is inserted upon
the ulna, while the tendon of the biceps passes between the ulna and the

radius ; if the hand be supinated vigorously and as far as possible, the attachment may be seen.

a. Carry the scalpel along the humerus between it and the brachialis so as to detach the muscle except at its origin.

§ 23. *Actions of Certain Muscles upon the Antibrachium.*—The anti-brachium is a LEVER divided at the elbow joint into a LONG ARM, and a SHORT ARM, the OLECRANON, Pl. I.

a With either of the muscles to be experimented with, biceps, tri-ceps and brachialis, the natural contraction may be imitated sufficiently by pulling proximad in the line of its length ; the effect is the same as if the muscle shortened itself like a piece of rubber that has been stretched.

§ 24. *Flexion.*—Grasp the humerus, the elbow down. Pull upon the brachialis. The hand rises and the arm is flexed just as when, in life, the muscle contracts.

The antibrachium constitutes a FLEXION LEVER, the power (brach-ialis) being applied between the weight (hand) and the fulcrum (elbow joint).

§ 25. *Extension.*—Grasp the humerus, the elbow up. Pull upon the triceps. The arm is extended just as when, in life, the triceps contracts. The antibrachium now constitutes an EXTENSION LEVER, the fulcrum (elbow joint) being between the weight (hand) and the power (triceps).

§ 26. *Rising.*—Grasp the humerus, the elbow down, it and the hand resting upon the table. Pull upon the triceps so as to straighten the arm at the elbow.

a. The condition is the same as when, upon all-fours, we rest the hand and antibrachium upon the floor and then raise the body by straight-ening the arm at the elbow. The weight (brachium, etc.,) is between the fulcrum (contact of the hand with the floor) and the power (triceps).

b. A more familiar example of this action is in rising upon the ball of the foot ; the weight (of the leg and body) is between the fulcrum (contact of the foot with the floor) and the power (muscles of the "calf") inserted at the heel. Illustrate this by means of the cat's leg ; rest the foot on the table and pull upon the *tendo Achillis* so as to raise the heel.

§ 27. *Antagonism of Flexors and Extensors.*—The biceps and brach-ialis and the triceps act upon the antibrachium in opposite directions ; they are OPPONENTS or ANTAGONISTS. If they should contract at once and with equal power there would be no movement ; the arm would sim-ply be "fixed" at the elbow ; illustrate this upon your own arm.

§ 28. *Reversal of the Fixed Point.*—In the previous observations the brachium is supposed to be fixed and the antibrachium movable ; this is the more frequent condition ; but in climbing the hand and antibrachium are fixed and the brachium is flexed upon the latter.

a. Illustrate this by tying a string tightly about the humerus and the proximal end of the brachialis ; then grasp the antibrachium and pull upon the muscle ; the brachium is flexed at the elbow. So far as the muscle is concerned the action is the same as in § 24, but the lo-cation of the fixed point is reversed.

§ 29. *Monarthal Muscles.*—The brachialis crosses but one joint (arthron) ; its attachments are upon two adjacent segments of the limb, and a single joint intervenes between the origin and the insertion ; hence it and similar muscles may be called MONARTHRAL.

§ 30. *Disarthral Muscles.*—But the biceps arises from the scapula
and is inserted upon the antibrachium ; hence between its attachments
intervene an entire segment (brachium) and two joints (shoulder and
elbow) ; it and similar muscles may therefore be called DISARTHRAL. The
monarthral were formerly called "short" and the disarthral "long."

§ 31. *Actions of the Biceps.*—Pull upon it when (*a*) the brachium is
fixed and show that it is then a flexor of the antibrachium ; (*b*) when the
antibrachium is fixed and show that it then a flexor of the brachium.
So far it acts like the brachialis.

But (*c*) grasp the arm at the elbow so that movement there is prevented and pull the biceps ; the scapula will be extended, *i. e.*, brought
more nearly into line with the brachium.   Ordinarily the scapula is fixed
and the whole arm extended at the elbow.   This may be illustrated by
resting one's arm upon the table, placing the fingers of the other hand
upon the biceps, and then lifting the arm as a whole ; *the biceps may be
felt to contract although no flexion occurs at the elbow.*

§ 32. *The Biceps as a Supinator.*—Pronate the hand as completely as
possible (Pl. 5, last paragraph of description) grasp the elbow and pull
the biceps strongly proximad ; the hand will be supinated partly so that
the pollex is above instead of at one side.

a.   Cut the brachialis entirely away ; pronate the hand again, noting
that the tendon of the biceps is drawn down between the ulna and the
radius ; when the muscle is pulled the tendon is drawn out again, and the
radius revolves upon its long axis.

§ 33.   Repeat the experiment, cutting away the remnants of other
muscles so as to ascertain that the tendon of the biceps is attached to a
tubercle (the bicipital tuberosity) of the radius which is visible only when
the hand is forcibly and completely supinated ; when the hand is pronated
the radius is rotated upon its own axis and the tendon is wound upon it
as is the rope upon the cylinder at an old fashioned well.   Now as pulling upon the rope would turn the cylinder in the opposite direction, so
pulling upon the tendon turns the radius and thus supinates the hand.

§ 34.   Observe the supinating action of the biceps upon your own
arm by placing the fingers over the muscle during vigorous supination ;
it will be felt to harden.

a.   Supination is employed not only in the actions named in the description of Pl. 5 but in turning the handle of a door, in using a sword,
etc.

§ 35. *Counteracting Muscles.*—The single muscle, biceps, may by its
contractions accomplish either of three different things, viz., (1) flex the
arm at the elbow ; (2) extend the whole limb at the shoulder ; (3) supi-
nate the hand.

a.   Either two, or even all three, of these may be performed at once,
but more commonly only one at a time.   The other two actions are then
prevented by the simultaneous contraction of the antagonizing muscles.

b.   For example, in supination, flexion at the elbow is prevented by
the triceps.   This may be observed upon the dissected arm of the cat, and
upon one's own as follows :

c.   Grasp the muscles of the brachium between the thumb and fingers
so that the latter are upon the triceps ; when the hand is supinated and yet
not bent at the elbow there will be felt a hardening of not only the
biceps but also of the triceps.

§ 36. *Ligamentous Action of the Biceps.*—Hold the arm by the elbow, the hand at the left, and the brachium at right angles with both the scapula and the antibrachium. Depress the scapula, *i. e.*, flex it upon the brachium ; the antibrachium will be flexed. Then extend the antibrachium and the scapula will rise to its former position. In all this the biceps acts just as if it were a ligament or a cord passing over a pulley at the shoulder.

a. During life the conditions are not quite identical because the muscle is slightly extensible and may contract considerably ; but one of the reasons for the difficulty we have in keeping the legs perfectly straight at the knees when we bend forward and try to touch the floor with the finger-tips, is that certain muscles of the thigh, attached to the leg below the knee, arise from the pelvis at such a place that when the hips are tilted as in stooping they are put upon the stretch ; their tendons (ham-strings) may be felt to become tense if the fingers are placed at the at the "inner" side of the knee.

§ 37. Transect the biceps at about the middle of its length and note the cut area ; the *power* of a muscle, *other things being equal,* depends upon the *thickness,* while its *length,* if the fascicles are parallel, determines the *distance through which it can contract.*

a. This general statement is more intelligible if we compare the contracting muscle to a strip of rubber that has been stretched ; the longer it is the greater will be its shortening ; the thicker it is the greater the force exerted.

b. The triceps is much larger in the cat and most quadrupeds than in man ; this is not so much for the forcible extension of the arm as such, but because this muscle also serves for the support and propulsion of the body.

§ 38. *The Ligaments at the Elbow.*—Trim off the triceps and other muscles that may remain about the elbow. Note that the arthral ends of the bones are enclosed in a CAPSULE as was the hip (Pract. I, § 5). If the humerus is turned from side to side there will be seen, at the ulnar side a thicker strip of the capsule extending from a prominence of the humerus obliquely distad to the ulna ; a similar one on the other side crosses the head of the radius. Cut away the rest of the capsule and note that these two LATERAL LIGAMENTS permit the usual flexion and extension, but prevent any considerable lateral movement ; if either is cut the bones separate readily at that side.

a. Divide both ligaments ; disconnect the humerus ; note the form of the apposed ends of the bones.

§ 39. *Relations of the Ulna and Radius.*—Separate the proximal ends of the two bones, and note the cylindrical form of the latter, turning as a pivot at the side of the former.

a. Disconnect the distal ends and note the reversal of the relative sizes of the two bones. The elbow joint is formed mainly by the ulna, the wrist by the radius.

§ 40. *Flexors and Extensors of the Hand and Fingers.*—Grasp the mass of muscles at the "inner" side of the antibrachium and pull toward the elbow ; the hand will be flexed at the wrist ; pull those on the "outer" side and it will be extended.

a. Observe the actions of the same groups of muscles upon your own arm ; in the cat as in man the flexors are more powerful ; indeed in most operations of the limb the act involves flexion, and extension merely returns the parts to a condition in which the act may be repeated.

§ 41. Slit the skin through the middle of the wrist, the palm and medius (middle finger). Dissect it up at both sides of the cut. Note the thickness of the fibrous and fatty pad in the palm and on the finger, constituting cushions, soft yet firm, warm yet not unwieldy. The special, conical pad at the wrist is supported by the pisiform bone (Pl. I).

§ 42. Cut off the palmar skin and pad. Note the absence of anything comparable with the mass of special short muscles constituting the human "ball of the thumb."

a. Also that, at the wrist, most of the muscles become continuous with tendons which run in a channel between the pisiform bone and the radius and are bound down by a fibrous band.

b. Slit this band. Lift the muscle and tendon and note the division of the latter into four for the four fingers, and their reinforcement by small muscles.

c. Extend the hand to the utmost and spread the fingers as much as possible. Hold the arm so that the hand cannot be flexed at the wrist. Then pull upon the muscle and note that the fingers are both flexed and drawn together.

§ 43. *Extension of the fingers.*—Dissect the skin and fascia from the dorsum. Several tendons will appear at the wrist. Disregarding those which stop there, isolate one along the middle of the limb which divides into branches, one for each finger.

a. Trace one to the last phalanx of the medius. Cut between the medius and the annularis and note the angles formed by the phalanges with one another.

b. Extend the medius and note that from the root of the claw there extends to the distal end of the first phalanx an elastic fibrous band ; this, without exertion upon the part of the animal, keeps the last phalanx retracted and the claw thus "sheathed."

PHYSIOLOGY PRACTICUMS.

PART II. THE THORAX AND ABDOMEN.

PRACTICUM III. THE THORAX OF THE CAT.

PLATES REQUIRED : I, II, IV, VI, VII.

§ 1. *Transecting the Neck.*—For this and the next practicum it will be convenient to separate the head from the trunk at about the point indicated in Fig. 2, corresponding nearly with the middle of the word NECK on Pl. II ; this will leave most of the neck with the head. Draw the head nearly into line with the trunk ; with a sharp arthrotome, by a steady circular incision, divide the soft parts as completely as possible to the bone ; let the cat be held with its back on a board : place in the incision the edge of a sharp hatchet or cleaver and strike it forcibly enough to cut the spine at one stroke.

§ 2. Examine the skeleton of man and the cat. Bear in mind that the components of any longitudinal series, as vertebræ, ribs or cartilages, are always numbered beginning with the most cephalic.

§ 3. Place the cat on its back, the neck to the left. As compared with Pl. 1, the ventral and dorsal regions are inverted.

§ 4. *Removal of the Extrinsic muscles.*—In doing this *take pains not to cut so deeply as to open the cavity of the thorax.*

a. Grasp the cut end of either PECTORAL MASS (Pl. IV. A—G) pull toward the meson, and trim off along the sternum.

b. The RECTUS is a ribbon-like muscle lying upon the cartilages just laterad of the sternum (Pl. IV). At about the middle of its length push the scalpel handle under it ; raise it a little ; transect it and dissect off the halves respectively cephalad and caudad.

c. Laterad of the cephalic part of the rectus are attached several muscles of the neck ; cut close to their attachments upon the ribs.

d. Farther dorsad are the remnants of the muscles referred to in Pr. I (§ 20) as suspending the thorax between the scapulas ; note that its thoracic attachments constitute more or less distinct *digitations*, like the teeth of a saw, whence the name SERRATUS. In removing it *cut with the scissors along the length of the ribs and cartilages,* that is in a general dorso-ventral direction, rather than lengthwise of the thorax as a whole.

e. On the caudal portion of the thorax remove all muscles for 1—2 cm. beyond the last ribs, but *do it very cautiously so as not to open the abdomen.*

§ 5 *The Spinal Muscles.*—Under this general title may be conveniently designated the thick mass lying along the dorsal part of the thorax (Fig. 1). Parts of them illustrate the way in which a slender tendon may be formed by the convergence of a thin FASCIA from the surface of a muscle. These muscles need not be removed.

§ 6. *On the cut end of the neck* note the two tubes, TRACHEA and ESOPHAGUS, the former ventrad and open, the latter dorsad and collapsed. Expose them as far as the first ribs by carefully cutting away the muscles and connective tissue, but *do not pull the soft parts cephalad lest the thorax be opened prematurely.* Save the larger VESSELS and NERVES so far as practicable. The JUGULAR VEIN has already been seen in Pl. III ; it may commonly be recognized from its location, its size, and from containing blood.

§ 7. *The Thoracic Framework.*—Manipulate the thorax so as to recognize by the touch the constituents of its framework. The SPINE is concealed by the massive SPINAL MUSCLES.

§ 8. *The Ribs.*—Count the RIBS on both sides, remembering that the last is short. The normal number is thirteen. Call the attention of the instructor to any anomaly in the number on either side.

§ 9. *The Costal Cartilages.*—Select the sixth rib, the middle of the normal series. Follow it to the sternum. At any spot between the sternum and the middle of the dorso-ventral diameter of the thorax the tracer-point is not resisted as in the part nearer the spine. This softer ventral part is the COSTAL CARTILAGE. Its place of junction with the rib may be determined either by testing with the point till the hard bone is reached, or by scraping off the adherent muscle remnants at about the place indicated in the figure. and noting the slight difference in color.

§ 10. *Sternal and Asternal Ribs.*—By manipulation determine that the first nine (rarely eight) cartilages reach the sternum ; the corresponding ribs are hence called TRUE or sternal, and the other four FALSE or asternal ; the 10th, 11th and 12th cartilages are attached, each to the one next cephalad, but the 13th lies free among the muscles and its rib is called FLOATING. Compare with the arrangement in man.

§ 11. *Intercostal Muscles.*—Between the adjacent ribs and cartilages are INTERCOSTAL MUSCLES ; their ectal layer has the same general direction as the ectal muscle of the abdomen ; indeed they may be regarded as parts of the same muscle, the thoracic portion interrupted at intervals by the ribs and cartilages.

§ 12. *The Sternum* (breastbone).—The cat's sternum is relatively longer, narrower and more flexible than the human ; it consists of eight or nine segments, the short one, bearing the letter M in Fig. 1, being sometimes wanting.

§ 13. *Opening the Right Thorax.*—This should be done *as directed* along the lines indicated in Fig. 1, and *without pulling upon the parts,* lest certain membranes be cut or torn.

a. With the scalpel divide the 6th cartilage near the rib, at about the point of crossing of the heavy interrupted line on Fig. 1 as represented by the arrow ; the complete division is indicated by the separability of the two parts.

b. Hold the scalpel with its edge upward, and nearly horizontal so as to form a slight angle with the thorax. Cautiously introduce the point, not more than 5 mm., and divide the intercostal muscle just cephalad. Then cut the next cartilage. and so continue till the second cartilage and first intercostal are divided.

c. From the starting point cut caudad including the eleventh cartilage ; then trim the first intercostal along the margin of the first cartilage, and with the scissors cut thence caudad, about 1 cm. from the sternum, including the eighth cartilage. ˙

d. Before continuing the incision lift the portion of parietes thus circumscribed, and note that the caudal end follows the curve of an ental attachment (to the diaphragm) ; with the scissors cut along that curve, at about 1 cm. from the attachment, to the point between the eleventh and twelfth cartilages where the second incision stopped. The direction is indicated approximately by the broken line in Fig. 1 crossing the 9th, 10th and 11th cartilages. Remove the piece, cartilages and intercostals, so freed.

§ 14. Cutting the Ribs.—The removal of the remaining portion of the parietes involves division of the ribs. For this the *rented practicum scissors must not be employed.* A strong pair is required and unless the student has them among his own instruments the instructor should be called upon. With old cats even nippers may be needed.

§ 15. Cut dorsad from the angle between the eleventh and twelfth cartilages (see Fig. 1) until the thick spinal muscles are reached, just caudad of the thirteenth rib. Then cut cephalad to the first rib and remove the other ribs and the intercostals.

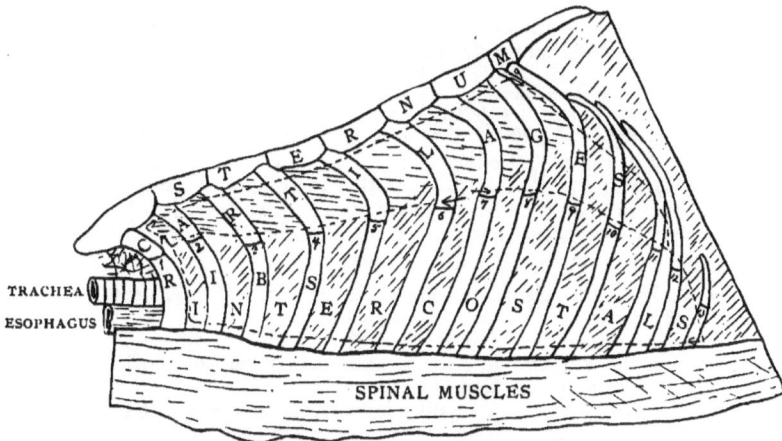

TRACHEA

ESOPHAGUS

SPINAL MUSCLES

FIG. 6. RIGHT SIDE OF THE CAT'S THORAX.

Diagram indicating the mode of dividing the right thoracic wall, while the cat is on its back, its head to the left. The letters of the word *sternum* are placed on the seven segments of the MESOSTERNUM, the PRESTERNUM and XIPHISTERNUM being left blank ; see Pl. I. The ribs are numbered 1—13, the numbers being placed just dorsad (under in the figure) of the lines of junction with their respective cartilages. The lines of incision are indicated by the interrupted lines, beginning at the 6th cartilage.

§ 16. Caution.—In the following examination *avoid displacing the viscera more than directed* and particularly all pulling or lifting of the sternum, on account of a delicate attachment to be described in § 29. Any liquid in the thorax may be sopped with absorbent cotton.

§ 17. *The Thoracic Parietes.*—Before displacing the contentsof the cavity thus opened to view, note the constitution of the walls or PARIETES. Dorsad are the ribs and intervening intercostals; ventrad, the cartilages and intercostals ; the ribs are supported from the spine, and the cartilages directly or indirectly join the sternum. Cephalad, the slight interval circumscribed by the spine, the sternum and the first pair of ribs and their cartilages, is wholly occupied by the trachea and esophagus, blood-vessels, nerves and connective tissue ; these will be seen later. The broad base of the thorax is formed by the DIAPHRAGM ; this will be more fully studied at Practicum IV ; at present there is to be seen only a narrow area of it under the overhanging strip of the lateral parietes.

§ 18. *The Pleura.*—Look at the ental surface of the removed lateral parietes and note that it is *smooth and shining ;* the cut edge may be separated as a delicate serosa, the PLEURA. If the overhanging strip of parietes left at the caudal margin of the opened thorax is reverted it will be seen that the pleura covering its ental surface is continued upon the diaphragm. Observe further that the visible surfaces of the exposed

viscera present the same appearance. Later it will be shown that the
serosa lining the walls, the PARIETAL PLEURA, and that which covers the
contained organs, the VISCERAL PLEURA, are *continuous with each other.*

§ 19. *The Lungs.*—There are visible three LOBES of the RIGHT LUNG
(Pl. VII), *cephalic, caudal* and *intermediate;* the last is as if wedged between
the other two, and does not, like them, reach the dorsal wall. The
HEART is the darker, rounded organ, near the sternum and partly covered
by the lungs.

§ 20. *Demonstrate the Lungs* by inserting into the trachea a blow-
pipe or a glass tube and inflating moderately ; when the inflation ceases
they collapse.

a. The lungs are in an unnatural condition because, for the sake of complete preser-
vation, alcohol was injected into them ; in a fresh state they would collapse much more
completely as soon as the thorax is opened ; see the lecture on Respiration.

§21. Lift the cephalic lobe of the lung, so as to expose the constricted neck by which it is attached ; note that the pleura is continuous over
its margin and upon the mesal surface, and that at the root it is reflected
in all directions ; it is thus continuous with the pleura which lines the
thoracic cavity.

§ 22. With the scissors amputate the cephalic and intermediate lobes
so as to leave a little stalk of each as in Pl. VII.

a. Compress either of them ; a frothy mixture of air and liquid will
escape from the cut end of a tube, BRONCHIOLUS, a subdivision of the
BRONCHUS or primary division of the trachea.

b. Inflate either lobe, holding it between the eye and the light ; at
the margin note the partitions between the ALVEOLI or air-sacks (sometimes called "air-cells"), the termination of the air-tubes, the smallest
subdivisions of the bronchioli.

c. Slit up a bronchiolus and note that its rather stiff walls have the
cartilaginous rings *complete*, unlike the trachea. The other tubes in the
lungs are branches of the PULMONARY ARTERY and VEINS.

§ 23. Lift the caudal lobe and note that the pleura is reflected from
not only its root, but part of its dorsal margin ; this margin is therefore
bound to the thoracic wall by a thin sheet of serosa which appears to be
single ; in reality it is double, since one layer comes from one side of the
lobe and one from the other ; Pl. VII, *Mpn.* Remove this lobe like the
others.

§ 24. *The Right Thoracic Cavity.*—The interior of the right half of
the thorax is now substantially as represented in Pl. VII ; Study this
plate and its description.

§ 25. *The Azygous Lobe.*—Besides the three lobes already examined
the right lung has a fourth or AZYGOUS LOBE, lodged in a sort of pocket
in the angle between the heart and the caudal lobe ; that it belongs to
the right lung may be determined by gently withdrawing it from the
pocket.

§ 26. *The Great Veins.*—At the margin of the pocket is a large vein,
probably containing blood ; this is the POSTCAVA (*vena cava inferior* or
"ascending cava") bringing blood from the abdominal viscera and legs to
the right auricle. The similar vein extending cephalad from the auricle
is the PRECAVA, *vena cava anterior* or "descending cava") bringing blood
from the head and arms. Joining the precava just cephalad of the
cephalic lung root is the RIGHT AZYGOUS VEIN. ′

a. Notwithstanding the identity of name there is no special relationship between the azygous vein and that lobe of the right lung. Moreover, neither of them is strictly azygous or mesal.

b. These large veins may not be recognized at first because they are collapsed and covered by pleura. Either of them is most easily demonstrated by pressing on the other two and carrying the fingers toward the heart.

§ 27. *The Phrenic Nerve.*—On the lateral aspect of the precava and postcava and the intervening portion of the auricle is a whitish cord, the RIGHT PHRENIC NERVE ; it comes indirectly from the myel (spinal cord) in the neck and is distributed to the daphragm ; see the lecture on Respiration.

§ 28. *The Thymus.*—Cephalad of the heart is a pale, lobulated mass, resembling a salivary gland ; this is the THYMUS BODY, larger in young cats, but in old ones sometimes insignificant. With the butchers both the thymus and the pancreas of the calf are sold as "sweet-breads."

§ 29. *The Thoracic Septum.*—Lift the sternum slightly and note in the interval between it and the heart and thymus a delicate, transparent membrane, the THORACIC SEPTUM. If no undue force has been used in preparing and opening the thorax it will form a continuous sheet, traversed by some vessels and nerves ; it will be more fully examined later, and is looked at now lest it be ruptured in opening the left thorax.

§ 30. *Opening the Left Thorax.*—Remove the lateral parietes of the left thorax as directed for the right (§ 13), taking especial care not to pull upon the sternum or cut too close to it. Lift the sternum and note the completeness yet transparency of the septum. It appears to be single, but is really double, the conditions being as follows : Each side of the thorax is lined by its own pleura, a closed sack. The thymus, heart and other mesal organs lie between the apposed mesal sides of the two pleural sacks ; but for a certain space near the sternum these apposed layers are in contact and apparently constitute a single membrane.

a. The independence of the right and left parts of the thoracic cavity provides that, in accident or disease. either side and its contained lung may be affected with less interference with the other.

§ 31. *The Left Lung.*—Note the incomplete separation of the cephalic and intermediate lobes of the left lung, and that there is no left azygous lobe ; the right may be seen through the pleura.

§ 32. *The Pulmonary Veins.*—If the lobes of the lung are displaced carefully laterad, at their root may sometimes be seen the PULMONARY VEINS full of blood.

§ 33. Amputate the lobes and observe the following parts. Their pleural covering may render their outlines indistinct.

a. The LEFT PHRENIC NERVE, crossing the septum at the pocket for the azygous lobe ; it is often bordered by a line of fat ; the cephalic part of its course may be less easily traced.

b. Near the diphragm, two cylinders, the ventral fleshy and pinkish, the ESOPHAGUS, already seen in the right thorax ; the dorsal, the AORTA, the great artery from the heart. The two following are not usually seen distinctly upon an alcoholic specimen.

c. The THORACIC DUCT, a corrugated tube.

d. The LEFT SYMPATHIC (sympathetic) NERVE, with its GANGLIA, the slight enlargements on the heads of the ribs.

The duct and nerve are better demonstrated upon freshly killed animals ; see *Anatomical Technology*, Figs. 103, 107, 108.

§ 34. Make a drawing of the left thorax similar to Pl. VII, but shade very lightly if at all.

§ 35. With the coarse scissors or nippers transect the sternum near the diaphragm ; also the first ribs ; then the intervening soft parts so as to remove the sternum. One or more vessels may be seen passing from the great thoracic vessels to the sternum. With the tracer tear the pleura and connective tissue so as to expose the aorta and esophagus more fully.

§ 36. Opposite the apex of the heart slit the aorta lengthwise and inflate it cephalad so as to demonstrate the following points :

a. The aorta extends cephalad and then turns somewhat sharply to the right and then caudad to join the base of the heart. Starting from the heart, therefore, the arch so formed is to the *left*, and the vessel itself is at the left rather than the right of the meson ; *Anatomical Technology*, Fig. 10r.

b. With all Mammals the aortic arch is left ; with all Birds it is right ; see the contrasted injected preparations in the Museum ; the matter is further discussed in the lectures.

c. From the aortic arch spring two great arterial trunks carrying blood to the head and arms.

d. If time permits expose them with the tracer ; the names and divisions are given in *Anatomical Technology*, Figs. 91, 101, 102 ; compare with the human arrangement as seen in the wall maps.

§ 37. Slit the aorta to near the diaphragm and note the orifices of the ten pairs of INTERCOSTAL ARTERIES.

§ 38. Cut the phrenic nerves, aorta and postcava about 2 cm. from the diaphragm ; then the precava, azygous vein and branches of the aortic arch. Pull all the parts cephalad, divide the pleural attachments with the scissors, and remove.

§ 39. *The Cat's Heart and Great Vessels.*—If time permits a hasty examination of these may be made as follows, but the beginner will find the sheep's heart more easy to dissect.

a. With the fingers tear off any fragments of fat or thymus ; then the esophagus ; then the trachea with its branches, the BRONCHI ; cut the bronchi near the lung-remnants. These latter are probably connected with the heart by the PULMONARY ARTERIES and VEINS. The veins may be full of blood.

b. Cut the pulmonary vessels close to the lung-remnants so as to free the latter.

§ 40. *Removal of the Pericardium.*—*a.* At about the middle of the entire length of the organ pinch up a fold of the membranous sack which incloses it ; slit the sack and remove, trimming quite closely to its attachments at the base, but without cutting the heart itself or the vessels.

a. The place and manner of attachment of the pericardium may be more easily observed in the sheep ; the object of this removal is to expose more fully the regions of the heart and the vessels.

§ 41. *Right Aspect of the Heart.*—*a.* Place the heart in the position shown in Pl. VII, the right side toward you, the apex toward your right.

b. Recognize the RIGHT AURICLE and VENTRICLE ; both are probably quite firm to the touch on account of the contained coagulated blood.

c. Note the attachments of the three great VEINS which bring blood to the auricle, the POSTCAVA and PRECAVA and the RIGHT AZYGOUS.

§ 42. *Left Aspect.*—Turn the heart upon its right side and see that the vessels lie in what seems to be their natural positions.

a. Recognize the LEFT AURICLE and VENTRICLE and the AORTA with its ARCH.

b. Note the two great BRANCHES from the aortic arch :

1 The smaller, farther from the heart, is the LEFT SUBCLAVIAN, carrying blood to the left arm.

2. The larger the BRACHIOCEPHALIC, supplies the head and right arm, whence its name.

c. Note the larger branches of the brachiocephalic ; commonly the first is the LEFT CAROTID carrying blood to that side of the head ; then the remainder divides into the RIGHT CAROTID and the RIGHT SUBCLAVIAN, this latter being continued as the AXILLARY seen in Pl. IV.

d. Sometimes the right subclavian arises as the first branch of the brachiocephalic and the remainder bifurcates into the two carotids ; *Anatomical Technology*, Fig. 4.

e. *The Pulmonary Vessels.*—Although these are cut short the PULMONARY VEINS may be traced to the left auricle, most easily by blowing, while the ARTERY passes under (caudad of) the aorta to the right ventricle.

f. Make an enlarged diagram of the heart and its vessels, indicating especially the branches of the aortic arch as described above. Compare with the condition in man in various mammals ; see Owen's *Comparative Anatomy*, III, Fig. 419.

§ 1. *Preparation of the Specimen.*—See Pract. III, § 38. A little cephalad of the dorsal attachment of the diaphragm, transect the spine, *etc.*, either summarily with a hatchet or cleaver as with the neck (Pract. III, § 1), or more instructively as follows :

a. Clear the soft part (excepting the aorta, postcava and esophagus) from the ventral side of the spine 3-5 cm. from the diaphragm.

b. Press the probe point against the spine at short intervals till it enters, indicating the location of an *intervertebral fibro-cartilage*.

c. Divide the *spinal muscles* at this level till bone is reached.

d. Push the arthrotome transversely into the cartilage and cut it so that the *vertebral centrums* separate slightly ; Plates I and VIII.

e. Flex the spine dorsad so as to increase the space ; note the *myel* and divide it with a scalpel.

f. Flex the spine still more dorsad from side to side, so as to indicate the location of the rather complex joints between the overlapping processes of the vertebræ. Use the arthrotome carefully so as to complete the separation.

g. When practicable adjacent students should have cats of opposite sexes.

§ 2. *The Diaphragm.*—The thoracic surface of the diaphragm (Pl. VIII) should present a marked convexity into the thorax and be approximately smooth ; if it is wrinkled, in the lateral parietes of the abdomen cut a small slit that may admit the blow-pipe, and inflate the abdomen till the diaphragm is as desired ; prevent the escape of the air by means of a compressor (spring clothes-pin or garment-clasp).

§ 3. Place the specimen on its back and steady it by means of the leaden cradle.

a. This is a piece of sheet lead about the size of this page (16×25 cm. 6.5×10 in.) bent across its length so as to form a W with rounded angles ; when inverted it makes a "cradle" which may be adjusted to the size of the cat and the position desired.

§ 4. Make a drawing of the diaphragm, *etc.*; consult Pl. VIII and its description ; but *draw the specimen and only the features it presents.*

a. The peripheral portion, next the spine, ribs and sternum, is muscular.

b. The fasciculi converge toward a non-muscular area, the CENTRAL TENDON ; this varies somewhat in size and form but is commonly crescentic, cordate or triradiate. It is sufficiently transparent for the abdominal viscera to be seen dimly through it.

c. The diaphragm is traversed by three tubes, already seen and drawn in Pract. III, *viz.*, the POSTCAVA, just ventrad of its middle, and probably full of blood ; the AORTA, near the spine and probably empty ; the ESOPHAGUS, fleshy, corrugated, nearly midway between the two. The phrenic nerves should be looked for ; if pulled gently cephalad there may be recognized some of their branches radiating in the muscular portion of the diaphragm. They should have been represented on Pl. VIII.

d. The cut edge of the PLEURA lining the other portions of the thoracic parietes should be represented by a continuous, sharp line. It is reflected upon the diaphragm and covers most of its surface ; but sinistrad of the POSTCAVA is an area corresponding to the location of the azygous or fourth lobe of the right lung, about which, under favorable conditions, may be traced the cut or torn edges of the pleura which was reflected to form the ventral and dorsal SEPTUMS (Pract. III, § 30).

c. If time and opportunity permit, compare with the human diaphragm as seen in the manikin, the wax model, and the preparations of young individuals.

§ 5. *Opening the Abdomen.*—Place the specimen on its right side. Compare with Plates II and IX. At about the middle of the abdomen (about the *N of ABDOMEN* in Pl. 2) pinch up a fold and slit for 2–3 cm. (about one inch). With the scissors cut thence, *without wounding or displacing* the viscera, to about 1 cm. of the meson ; thence cephalad to about the same distance of the diaphragm ; thence along it to near the spinal muscles. Return to the meson and cut thence to the pelvis and dorsad to the spinal muscles. The flap of abdominal parietes may then be turned dorsad.

a. The best instrument for dividing the abdominal wall without cutting the viscera is a probe-pointed bistoury.

§ 6. *The Peritoneum.*—Note that the ental surface consists of a smooth serosa, the PERITONEUM ; also that there are recognizable at least two layers of muscle, the fibers of the ectal having the direction indicated in Pl. 2, those of the next layer extending caudo-laterad at an acute angle with them, *i. e.*, nearly parallel with the fibers of the Latissimus and the great skin muscle ; Pl. II. By an incision with scissors along the lateral border of the spinal muscles the flap of parietes may be removed as in Pl. IX.

§ 7. *General View of the Viscera.*—Compare the specimen with Pl. IX ; identify organs and features as far as possible without dislocating them permanently.

§ 8. *The Omentum*—(known among the butchers as the "caul")—is a membranous "apron" spread more or less completely over the abdominal viscera ; sometimes it has been displaced in preparing the specimen, but usually it is recognized without difficulty ; it is traversed by blood-vessels, and in well-nourished animals is commonly streaked with fat ; through the transparent areas between the fat masses may be dimly seen the folds of the intestine.

a. Make a small slit in the omentum and blow at the cut edge until the air enters between two layers and inflates the omentum as a loose bag. In reality each of the two layers thus separated consists of two closely united layers of the peritoneum reflected from the viscera ; for details see the manuals of Human Anatomy and *Anatomical Technology*, pp. 278, 280.

b. Lift the caudal free border of the omentum and gently draw it sinistro-cephalad, taking care not to displace the viscera ; with the scissors cut it off and remove.

§ 9. *The Umbilicus.*—From the LIVER extends caudad at the meson a fold of peritoneum ; a similar one extends cephalad from the BLADDER ; in young cats, at some point between the ends of these two folds, look for a mesal spot differing more or less in color or texture from the surrounding parts ; this is the UMBILICUS or navel, the place of attachment of the FUNIS or umbilical cord by which the unborn kitten is connected with the mother.

a. Before and shortly after birth, could have been traced to it from the liver a vein of which the maternal blood reached the fetus, and from the bladder a pair of arteries carrying the fetal blood in the opposite direction ; also a hollow cord, the URACHUS, continuous with the bladder itself. The relations and significance of these parts are presented in the larger works on Anatomy and Physiology and in one of the Lectures. The human umbilicus is obvious through life from the ectal surface also.

§ 10. Place the specimen on its back. Look at the prepared skeleton or at the diagram for the XIPHISTERNUM, the caudal extension of the sternum ; it may be felt at the meson, opposite the lobe of the liver marked *I*, especially if the cut end of the sternum be moved a little ; cut away the muscles carefully so as to expose the xiphisternum, noting its peculiar shape and that the caudal third or fourth is cartilage. The same part in man is shorter and less shapely.

§ 11. Remove the remaining abdominal parietes. From Pl. IX recognize as many as possible of the viscera by their colors, textures and relative positions ; note the closeness with which they are packed, their overlappings and the smoothness of their surfaces, covered by peritoneum.

§ 12. *The Mesentery.*—The middle and caudal portion of the abdomen is chiefly occupied by the irregular coils of the SMALL INTESTINE. Lift a loop of it and note that from one side extends a membrane connecting the intestine with the adjacent parts of the intestine and with the dorsal wall of the abdomen ; this is the MESENTERY. With the tracer, close to the intestine, tear the membrane slightly and note that between it and the corresponding membrane from the opposite side is a little interval ; at or near the dorsal attachment of the mesentery the two layers again diverge, passing to the right and left respectively to become continuous with the general peritoneal lining of the abdomen. From this it may be seen that the intestine lies between two layers of serosa, the adhesion of which constitutes the mesentery ; see *Anatomical Technology*, Fig. 78 ; compare with the thoracic septum, Plate VIII, and Pract. III, § 30.

§ 13. *The Mesenteric Vessels.*—The mesentery is traversed by three kinds of vessels connected with the intestine :—(*a*) ARTERIES, carrying blood to it ; (*d*) VEINS, bringing from it blood which enters the PORTAL VEIN at the liver ; (*c*) LACTEALS, bringing chyle to the thoracic duct. The lacteals are probably invisible in ordinary alcoholic specimens, but may be seen in the special preparations and in animals killed shortly after taking fatty food ; *Anatomical Technology*, Fig. 103. Near the dorsal attachment of the mesentery may be seen one or two MESENTERIC GLANDS, which, like lacteals, are parts of the general LYMPHATIC SYSTEM.

§ 14. *The Stomach and Small Intestine.*—Through the esophagus inflate the STOMACH, first completely to display its possible size, then moderately, and tie the esophagus. Displace the adjacent parts so as to expose the entire stomach. The esophagus opens into a larger, subglobular portion, more at the left, nearer the heart, and called the CARDIAC region ; the more slender PYLORIC portion is flexed quite sharply on the other, and is continuous with the DUODENUM or first part of the small intestine. The PYLORUS, constituting the boundary between stomach and intestine, may be recognized by manipulation as a distinct thickening of the muscular coat ; § 22 *c*, and *Anatomical Technology*, Fig 79.

a. The three commonly accepted divisions of the small intestine, DUODENUM, JEJUNUM and ILEUM, are not sharply defined although, in the cat, the first may well be regarded as coëxtensive with the attachment of the pancreas (Pl. VI, D.

§ 15. *The Spleen.*—This is a dark red organ lying sinistro-caudad of the cardiac region of the stomach ; it is elongated and but loosely connected to the stomach by a fold of peritoneum. In man the spleen is massive and its attachment is much more close.

a. The cat's spleen varies considerably in size, but is never so compact as in man.

b. The spleen is abundantly supplied with arteries and veins, but has no cavity or excretory duct ; it is hence sometimes called a *ductless gland*.

§ 16. *The Pancreas.*—This is a pale, lobulated organ attached quite closely along the duodenum, and sending an additional portion toward the spleen ; it resembles the thymus (Pract. III, § 28), or a salivary gland Pract. VI, § 3). The general location of its two ducts, by which the pancreatic liquid is poured into the duodenum, may be recognized from the closer adhesions, but there may be neither time nor light for tracing them in detail ; see *Anatomical Technology*, Figs. 79 and 81.

   a. The pancreas is naturally pale but may be discolored and dark in the alcoholic specimen. The portion extending toward the spleen is but slightly represented in man.

§ 17. *The Liver.*—This is more nearly mesal and symmetrical than in man, and more completely subdivided into lobes, permitting greater flexibility of the animal. Note the reflections of its peritoneum upon the diaphragm.

§ 18. *Removal of the Diaphragm.*—With the scissors cut the diaphragm from all its attachments, (*a*) at the periphery ; (*b*) at the peritoneal reflection upon the liver ; (*c*) where it is traversed by the esophagus, aorta and postcava but without cutting these three tubes.

§ 19. Tilt the ventral margin of the liver cephalad so as to expose more of the intestine. Note the continuity of the small intestine with the stomach cephalad and with the large intestine (colon); also that the whole length is quite closely connected by mesentery excepting the caudal part of the colon.

§ 20. *Detaching the Small Intestine.*—At any point not less than 10 cm. (4 in.) from the cecum cut the mesentery close to the intestine ; continue to cut, first one way and then the other, until within about 5 cm. of the cecum and about 10 cm. of the stomach. At each of these points compress the intestine so as to force the contents either way for 1–2 cm. ; tie firmly with a cord at both ends of the vacated space and cut between. Ask the janitor to remove the detached portion of intestine.

§ 21. *The Gall-Bladder or Cholecyst.*—This is lodged in a cleft of one lobe of the liver, Pl. X, 2, 3 ; if full, compress it steadily until the bile passes through the contorted BILE-DUCT into the duodenum near the attachment of the pancreas. If it is empty slit it and inflate, preventing the escape of air with the fingers ; the air may be made to pass through the duct into the intestine.

   a Remember that the gall-bladder is not a source of bile, but merely a *reservoir ;* as the bile comes from the substance of the liver through the hepatic ducts a portion of it flows back to the gall-bladder and is there stored up. See Martin, Fig. 52 and *Anatomical Technology*, Fig. 79. Remove the liver, pancreas and spleen.

NOTE.—The directions in the following paragraph may be disregarded if time is lacking or strong objections are felt to following them :

§ 22. *Opening the Stomach.*—Cut the stomach away together with the attached portions of the esophagus and small intestine. Take it to the sink. Over the water-pail slit with the scissors along the greater curvature from the orifice of the esophagus to the end of the duodenum ; press out the contents ; rinse off the mucosa with a stream of water, gently rubbing the surfaces together. When cleansed from food and mucus take the stomach to the table and note the following points :

   a. The gastric mucosa is loosely adherent to the muscular coat.

   b. It may present several RUGÆ, folds or ridges, which permit considerable distention of the organ without undue tension of the mucosa. The rugæ are usually apparent in fresh specimens, but may be obliterated by distention of the stomach as by the injection of alcohol.

   c. The pylorus is indicated, not by a valve in the ordinary sense, but by an ental, annular ridge consisting of a thickening of the muscular coat and constituting what is called a SPHINCTER.

d. The intestinal mucosa lacks the VALVULÆ CONNIVENTES or transverse ridges that exist in man.

e. Although the *valvulae conniventes* are absent from the cat's intestine a good idea of their nature as folds of the mucosa may be gained from the rugæ of the stomach, although the latter are only temporary.

f. The velvety feel of the intestinal mucosa is due to the VILLI ; these may be seen with a lens.

g. Grasp the bile-duct and look for the orifice by which it enters the duodenum.

§ 23. *The Large Intestine.*—The longer, intermediate portion of the large intestine is the COLON, *cl.;* the three portions, *"ascending," "transverse"* and *"descending,"* are less easily distinguished than in man ; the caudal portion, lodged mostly in the pelvis and terminating at the ANUS, is nearly straight and is called the RECTUM.

a. Bi-ligate the rectum (§ 20) and remove the rest of the intestine. If desired the cecum with the adjoining portions of the colon and ileum may be opened as with the stomach (§ 22) so as to show the sphincter which guards the ileo-cecal orifice (*Anatomical Technology*, Fig. 80) ; in man there is a true valvular arrangement.

b. The regurgitation of the contents of the colon into the small intestine is further guarded against by the projection of the sphincter into the former.

§ 24. *The Cecum.*—As shown in Pl. X this is a short part of the large intestine projecting beyond the point of continuity with the small. Note the absence of the slender appendix of the human cecum.

§ 25. *The Kidneys.*—The LEFT KIDNEY is partly shown in Pl. X, more fully in Pl. XI. Both have been more fully exposed by the removal of the other viscera. They are commonly overlaid in part by fat, but it can probably be seen that the right lies a little farther cephalad, the reverse of the human condition. Kidneys and fat are covered by the peritoneum, and are thus, strictly speaking, ectad of the true abdominal cavity ; *Anatomical Technology*, Fig. 78.

§ 26. *The Adrenal.*—Cephalo-mesad of each kidney is a pale, lobulated body, like a small pancreas ; this is the ADRENAL or "supra-renal capsule," one of the ductless glands ; § 15, *b.* The left is shown in Pl. X, A. In man the adrenals are closely attached to the kidneys and shaped like cocked hats.

§ 27. From the left kidney dissect off the peritoneum and fat, beginning at its lateral margin, and lift it slightly from its bed. From the concave mesal side will be seen to pass mesad an ARTERY, a branch of the aorta, and a VEIN from the postcava. Also, meso-caudad, extends the URETER, a slender tube imbedded in fat, Pl. XI ; if the kidney is drawn cephalad a little the ureter will be put upon the stretch so as to be more easily recognized.

§ 28. *The Urinary Bladder.*—If the bladder is drawn out of the pelvis and turned caudad (Pl. XI,) it will be seen to have a narrow *neck;* the ureters enter the dorsal side of the neck. If the bladder still contains liquid, steady pressure may expel it through the URETHRA and ectal organs ; if empty, air may be introduced through a slit as with the gall-bladder (§ 21).

§ 29. *Dissection of the Kidney.*—Slit the left kidney lengthwise from its convex border so as to open its cavity, the PELVIS (not to be confounded with the region of the trunk which has the same name) ; if air is blown into this cavity toward the ureter, it will sometimes traverse the latter and enter the bladder so as to inflate it.

a. In Pl. XI the plane of section is nearly equidistant between the dorsal and the ventral surfaces of the organ ; but if the section is made nearer either of these surfaces the cavity will be larger and there will be four pyramids resembling in outline the single papilla which they converge to form ; *Anatomical Technology,* Fig. 86. In the human kidney there are several papillæ.

§ 30. Transect the right kidney so as to gain an idea of the relative positions and colors of the two portions, the ectal or cortical portion consisting largely of the *glomeruli,* and the ental or medullary part consisting largely of *tubules* ; Pl. XI.

a. A section of the kidney, especially when fresh, seems to indicate the existence of a third or intermediate zone ; it was particularly distinct in the specimen from which Fig. 19 was made. Injected preparations show that it is a vascular portion of the medullary zone.

§ 31. *The Uterus.*—In the female, between the bladder and the rectum is a hollow, fleshy organ, the UTERUS or womb, Pl. XI ; it is Y-shaped, the mesal stem caudad, the two branches extending cephalad toward the kidneys.

a. Each branch consists of two parts, a larger, the CORNU or horn, and a smaller, farther cephalad, the OVIDUCT or Fallopian tube.

b. With the cat, dog, pig and many other mammals the young are developed in these lateral portions, but in the human species development occurs normally in the mesal part or "body" of the uterus, and the branches are relatively insignificant.

§ 32. *The Ovary.*—Near the kidney is the OVARY, an oval, lobulated body. Slit the uterus, pass the probe either way and endeavor to uncoil the oviduct and find its orifice near the ovary. This is a difficult task and hardly to be accomplished at an ordinary practicum.

§ 33. *The Spermaduct.*—In the male, each ureter is crossed by a tube, the SPERMADUCT or *vas deferens.* Traced in one direction it will be found to enter the neck of the bladder farther caudad and nearer the meson than the ureter ; see Mivart, Fig. 115. In the other it will be found attached to the TESTIS or testicle. As its name implies, the spermaduct is the correlative of the oviduct of the female, and serves to carry to the emittent organ the SEMEN (seminal liquid or sperm) for the fertilization of the ovum.

PRACTICUM V. THE SHEEP'S HEART.

PLATES REQUIRED : XII, XIII, XIV.

§ 1. The heart has been prepared by filling with alcohol after the removal of the blood. The vessels are cut short. Bits of the lungs are left attached. The pericardium has sometimes been mutilated by the butcher; when it remains there are usually attached bits of fat which may be torn off with the fingers.

§ 2. *Removal of the Pericardium.*—At about the middle of the length of the heart slit the pericardium and "girdle" it completely so as to separate the apical (caudal) from the basal (cephalic) portion. Remove the former; note the smoothness of the ental surface. It and the apposed ectal surface of the heart are covered by serosa and are moistened during life by the secreted serum.

a. Turn the basal portion cephalad, inside out, like the finger of a glove. At varying distances from the base it is attached to the heart and vessels, and the ental serosa is reflected thereon; recall the relations of the two layers of the thoracic serosa, the pleura, Practicum III, § 21 ; the cardiac serosa in like manner consists of a VISCERAL LAYER, the EPICARDIUM, covering the heart, and a PARIETAL LAYER, lining the pericardium, and the two are continuous near the base of the organ ; see Figs. 9 and 10, the PERICARDIAL LINE.

b. Trim the pericardium along or near the line of attachment.

§ 3. *General Topography of the Heart.*—Recognize the several regions by comparison with that of the cat and Pl. XII, and by the following features :

a. The APEX or caudal end is regular, conical, smooth, firm and fleshy ; it is formed by the muscular VENTRICLES.

b. The BASE or cephalic end is irregular and wider, and presents not only the thin-walled AURICLES, but also vessels and fat and remnants of lungs.

c. At about the middle of the length, one diameter, the dextro-sinistral, is decidedly the greater, so that the heart is depressed, *i. e.*, as if from dorso-ventral pressure.

d. Of the two wider aspects, the convex is ventral, while the dorsal is, as a whole, concave. Of the two narrower sides, the right is the more convex.

e. Make sure of these aspects, so as to recognize them with the eyes closed.

f. Remember always that *right* and *left* and other descriptive terms apply to certain regions of the heart without reference to the right and left of the observer and irrespective of the way in which the organ is held. For example on Pl. XII the right of the heart corresponds with the observer's in the lower figure, but in the upper it is at his left.

§ 4. *The Vessels.*—These may be recognized from the cat and from Figs. 9 and 10. Cut off the tied ends. Note the following points :

a. The arteries, AORTA (*A*), its CHIEF BRANCH (*B*), and the PULMONARY ARTERY, maintain a cylindrical form, and their cut ends are naturally circular ; the great VEINS, POSTCAVA and PRECAVA, have thinner walls in proportion to their size and collapse more or less completely.

b. The POSTCAVA forms nearly a right angle with the long axis of the heart, indicating that in the sheep the heart lies more obliquely than in the cat.

c. The right AZYGOUS VEIN, which in the cat (and man) opens into the precava near its root, is rudimentary and may not be found. But a LEFT AZYGOUS should be looked for as represented in Pl. XII, *Az.;* the peripheral end is probably tied ; it will be more fully seen upon dissection.

d. On the ventral aspect, between the two auricles, is the prominent PULMONARY ARTERY, extending from the base of the RIGHT VENTRICLE obliquely sinistro-cephalad. The part of the ventricle from which the artery springs is the CONUS.

e. The AORTA, with its principal branch (B) will be seen more distinctly at a later stage.

§ 5. *The Sulcus Terminalis.*—The right auricle presents two regions, one smooth, between the attachments of the postcava and precava, the other corrugated slightly, extending dextrad toward the pulmonary artery. Between the two areas is a slight furrow which begins at the notch at the dextro-cephalic corner of the auricle and crosses diagonally to the sinistro-caudal angle. This furrow, more distinct in man, is the *sulcus terminalis* of His ; the two regions of the auricle are the ATRIUM and the APPENDIX.

§ 6. *Dissection of the Heart.*—The order of dissection follows the course of the blood through the organ. Bear in mind that, although anatomically united, and acting together as muscles, *as to their cavities, the right and left sides of the heart are entirely separated by complete partitions between the two auricles and between the two ventricles.* Physiologically, between the right ventricle and the left auricle, intervene the lungs ; in like manner between the left ventricle and the right auricle, intervene all the other parts of the body, including the substance of the heart itself ; see *Anatomical Technology*, Fig. 92.

a. If the directions are followed closely and carefully, the dissected heart may be worth keeping as a guide in future dissections.

§ 7. *Opening the Right Auricle.*—*a.* Hold the heart with its ventral side toward you, as on Pl. XII. At the dextro-cephalic angle of the appendix, 5–10 mm. from the margin, push in a scissors-point. Cut thence caudad and then sinistrad, keeping at about the same distance from the margin, to a point about midway between the right margin and the pericardial line where it crosses the postcava.

b. Lift the edge of the flap and note the wide mouths of the POSTCAVA and PRECAVA, separated by a prominent ridge ; Pl. XIII, 9.

c. Cut carefully almost to the pericardial line ; thence, clearing the ridge just mentioned and another at the opposite side of the precaval orifice, cut to the point of departure.

§ 8. *The Coronary Sinus.*—The large orifice of this is seen just caudad of that of the postcava, Pl. XIII. Slit its ectal wall for 1–2 cm.

a. Extending obliquely cephalad from the sinus is the (left) AZYGOUS VEIN ; with the scissors slit it to its cut (or tied) end (Pl. XII, Az.), noting its semi-circular course.

b. Near the angle of the appendical part of the left auricle is the large orifice of the CARDIAC (coronary) VEIN, *bringing blood from the substance of the heart ;* it is seen transected in Pl. XIV.

c. Between the cardiac vein and the orifice of the sinus are several small openings, the FORAMINA OF THEBESIUS, the mouths of small veins.

d. With the scissors remove the free portions of the walls of the atrium, postcava and precava. Compare Pl. XIII ; note one or two Thebesian foramina where the precava becomes continuous with the atrium.

§ 9. *Opening the Right Ventricle.*—This must be done very carefully, especially if a permanent preparation is to be made.

a. On the ventral aspect of the heart (Pl. XII) note the line of attachment of the pulmonary artery, the heart being darker and firmer than the vessel.   This part of the ventricle is sometimes distinguished as the CONUS.

b. At a point on the ventricle about 1 cm. from the artery and in the line of its right border, insert the scalpel-point not more than 1 cm. and cut sinistrad, at a right angle with the axis of the vessel, for its width, stopping 5–10 mm. from the interventricular furrow.

c. Then, inserting the scalpel no deeper, cut parallel with the furrow until the right margin of the ventricle is reached near the apex of the heart.

d. Lift the free corner of the triangular flap so as to get a partial view of the cavity of the ventricle ; look particularly for a cylindrical band connecting the interventricular septum with the lateral wall.

e. With the scissors cut from the first point of departure on the hypothenuse of the triangle, swerving a little to the left if necessary to avoid the attachment of the band just mentioned ; this may now be seen distinctly ; it is the MODERATOR BAND, constant (though varying in size) in the sheep and some other animals, but infrequent in man.   It is supposed to limit the distention of the ventricle.

f. Keep the cut edges of the ventricular wall apart with large pins or with the fingers and sketch the cavity with special reference to the moderator band.

§ 10.   From the caudal or apical end of the triangular opening already made, cut along the left border of the ventricle to within about 2 cm. of the auriculo-ventricular line.

a. Lift the triangular flap so formed and note that from certain parts of its ental surface, in addition to the moderator band, spring muscular columns (COLUMNÆ CARNEÆ) which extend cephalad and are connected with fibrous cords (CHORDÆ TENDINEÆ).

b. From the apex and margins of the flap trim off so much as has no direct connection with the band or columns.

§ 11.   *The Tricuspid Valves.*—Pass the finger from the auricle into the ventricle and distend the AURICULO-VENTRICULAR ORIFICE.   Note that it is surrounded by three fibrous sheets which hang down into the ventricle and are connected at the sides with the cords and columns above mentioned.   These are TRICUSPID VALVES or RIGHT AURICULO-VENTRIC-ULAR VALVES ; one lies against the ventricular septum, the others re-spectively near the right and left sides of the ventricle.

a.   The cords are attached mainly at the sides of the valves, but from the middle of the free border of the septal valve there sometimes pass several cords to a depression in the septum, the columns being wholly short or wholly absent.

b.   Some of the cords from the right of the septal valve may spring from the septal end of the moderator band.

§ 12.   The parts may be exposed more fully by removing the rest of the lateral wall of the ventricle, a patch being retained, if desired, for the · attachment of the moderator band.   A still better view may be had by cutting the auriculo-ventricular "ring" with the scissors between the two valves that meet at the right.

a.   The valves permit the ready passage of blood from the auricle to the ventricle but a reflux is checked by the crowding of the valves towards a common point, by a certain portion of blood getting behind them, just as swinging doors may be closed by the pressure behind them of a few individuals, although the crowd as a whole is striving to pass through.   The columns and cords prevent the free edges of the valves from being carried too far.

b.   In studying the action of the valves on an alinjected heart it should be borne in mind that the distention of the right ventricle by the alcohol is sometimes excessive, and may prevent the complete closure of the orifice which occurs in life.

§ 13.   *The Pulmonary Artery.*—About midway between the ventricle and the pericardial line cut a window in it at least as large as the triangle indicated on Pl. XII.

a. Admitting the light, look toward the ventricle and note that the mouth of the artery is surrounded by three membranous SEMILUNAR (or sigmoid) VALVES.

b. Pass the scissors point from the ventricle into the artery and cut toward the middle of the window already made. On divaricating the sides of the artery it will be found that (1) one of the valves has been medisected ; (2) each constitutes a sort of pocket ; and (3) opposite each there is in the wall a depression, a SINUS OF VALSALVA.

c. The semilunar valves permit the blood to pass freely from the ventricle into the artery, but a reflux is prevented by some of the blood getting behind the valves and closing them ; without the sinuses the valves might be pressed so closely against the walls as to prevent their closing.

§ 14. *Opening the Left Auricle.*—Note the two portions, the atrium dorsad, crossed by the azygous vein ; the appendix ventrad. Cut a window in the appendix so as to expose its entire cavity.

a. Note the general resemblance to the right appendix, but the greater prominence of the trabeculæ.

b. This prominence of the trabeculæ and the concomitant depth of the interspaces gives to the wall a spongy character which is even more marked in all parts of the heart in the human fetus, and is permanent with some lower animals, *e. g.*, the turtle.

c. Between the appendix and the atrium, toward the base of the heart, is a prominent RIDGE, coinciding approximately with the azygous vein.

§ 15. *The Pulmonary Veins.*—Hold the heart so as to see the depth of the atrium and note that it presents a RIDGE at right angles with the one just mentioned.

a. At·either side of this ridge is a cavity into which open the PUL-MONARY VEINS of that side ; Pl. XIV.

b. From the appendix cut through the ridge and azygous in two lines converging in the cavity for the left pulmonary veins.

c. Follow up the incision to the end of a vein which is tied ; this is the one through which the left side of the heart was injected. The others are probably tied up in the roots of the lungs.

§ 16. *The Fossa Ovalis.*—Hold the heart so that the auricular septum is between the eye and light. Near the ridge between the pulmonary veins is a depressed translucent area about one cm. in diameter. It is thin and yields to pressure. In the right atrium it is at the orifice of the postcava. The depression is the FOSSA OVALIS ; Pl. XIV.

a. In the fetus there was here an orifice, the *foramen ovale*, between the two auricles so that the blood from the right passed into the left instead of into the right ventricle ; at or soon after birth it should close. Sometimes, especially in young individuals, there persists a slit-like remnant of the orifice to which the attention of the instructor should be called.

b. Draw the left auricle, locating the fossa ovalis.

§ 17. *Opening the Left Ventricle.*—With the scalpel transect the heart just caudad of the septal attachment of the moderator band.

a. The cavity thus exposed is the LEFT VENTRICLE.

b. Make an outline of the cut end of the apical piece ; supply the wall of the left ventricle from memory.

c. The much greater thickness of the left walls is required in order to force the blood through the arteries of the entire body.

d. Slit the septum from the cut surface to the apex of the left ventricle ; note that at the tip the wall is no thicker than that of the right, illustrating the resisting power of living muscle.

e. The left ventricle forms the apex of the heart, and the right is as it were laid upon it.

§ 18. With the scissors cut from the auricle through the ventricular wall near the septum ; then divaricate the sides.

a. The auriculo-ventricular orifice is surrounded by what seems at first a continuous VALVE ; this, may, however, be divided somewhat arbitrarily into two parts, a SEPTAL, applied against the septum and a LATERAL, opposite it ; hence they are called BICUSPID or MITRAL.

b. Note the CHORDÆ TENDINEÆ and COLUMNÆ CARNEÆ as in the right ventricle ; Pl. XIV.

c. Remove part of the lateral wall of the ventricle and note on the cut surfaces the ends of the CARDIAC VESSELS as in Pl. XIV.

§ 19. *The Aorta.*—Trim off the remnants of the right auricle and ventricle. Slit the septal bicuspid valve for half its length, and note that it covered a large orifice, the MOUTH OF THE AORTA, surrounded by three SEMILUNAR VALVES.

a. Pass a flexible probe thence out of the cut end of the aorta or its branch, or make two stiff probes meet half way.

b. Remove the lateral wall of the aorta and its branch to near the level of the pulmonary valves.

§ 20. *The Ductus Arteriosus.*—Where the aorta crosses the pulmonary artery look for a slight depression or foramen, the aortic end of the DUCTUS ARTERIOSUS, through which, in the fetus, the blood from the right ventricle entered the aorta from the pulmonary artery at the point so marked (obscurely) in Pl. 12. Tear the two vessels apart carefully and look for the remnants of the tube. If it is found pervious the instructor's attention should be called to it.

§ 21. *The Aortic Arch.*—These resemble the semilunar valves already seen in the pulmonary artery. One is ventral, the other two right and left.

a. Pass a scissors-blade between the ventral and right valves ; cut the ventricular septum and divaricate the sides.

b. The right sinus of Valsalva resembles those in the pulmonary artery.

§ 22. *The Cardiac Arteries.*—The left and ventral sinuses of valsalva present each a circular orifice the adit of a cardiac or coronary artery.

a. These arteries supply the substance of the heart with blood. The corresponding veins were mentioned in § 8. These arteries and veins are the intrinsic blood-vessels of the heart.

b. Commonly there is one arterial orifice at each sinus ; sometimes two or even three.

c. The arteries may be traced for a short distance only in the dissected specimen.

§ 23. If time permits make drawings of various aspects of the dissection.

دا

36

PHYSIOLOGY PRACTICUMS.

PART III.   THE HEAD AND ORGANS OF SENSE.

PRACTICUM VI.   THE HEAD AND NECK OF THE CAT.

PLATES REQUIRED, I, II, XV, XVI, XVII.

§ 1.   If the skin of the head has been retained, remove it from the left side as indicated in Pl. XV, beginning with the cut edge on the neck.   Grasp that between the left thumb and finger and dissect it up for a short distance ; then divide the skin with the scissors along the dorsal and ventral lines indicated.   Lift the flap as before and continue till the area is exposed.

  a.   Unless the student has plenty of time the removal of the skin should be done in advance by an assistant.

  b.   Compare with Pl. XV.   Bear in mind that the figure shows several parts which are to be disregarded at present ; that no two individuals are absolutely identical ; that the parts are probably covered by a fibrous sheet or FASCIA so as to be less distinctly visible than on the figure.

§ 2.   *Removal of the Fascia.*—With the forceps lift the fascia at any point and try to tear it off with the tracer ; if compelled to use the scalpel or scissors, be very careful not to cut anything but the fascia.   The JUGULAR VEIN can probably be recognized from containing some blood.

§ 3.   *The Parotid Gland.*—So much of the PAROTID GLAND as may remain lies dorso-cephalad of the U-shaped loop of the jugular.   Along a line between the parotid and the angle of the mouth look carefully for the DUCT OF STENO, the excretory duct of that gland ; there are two or three nerves that might be mistaken for the duct, but they are solid and do not join the gland by three or four roots.   At the distance of about 1 cm. from the angle it pierces the cheek to open into the mouth opposite the largest of the teeth in the maxilla (upper jaw) the third from the canine or eye tooth ; see Pl. XVII, where a bristle is in the duct.

  a.   With the scissors cut through the cheek near the mandible ; reflect the flap dorsad ; pull upon the duct gently and this will indicate the location of the papilla through which it opens.

§ 4.   *The Submaxillary Gland.*—Dissect off the parotid ; this will more fully expose the SUBMAXILLARY GLAND which lies in the loop of the jugular vein ; its duct opens in the floor of the mouth (See Pl. XVII, *Ductus Wharton*) but time will not permit the examination of it, or of the sublingual and molar glands ; *Anatomical Technology*, 302.   Just cephalad of the submaxillary and separated from it by the jugular are one or two LYMPHATIC GLANDS.

§ 5.   *The Eyelids.*—Slit the skin at the lateral angle of the eye so as to permit the wide separation of the lids.   Note (*a*) the large size of the

EYEBALL ; (*b*)the UPPER and LOWER LIDS, hairy on the ectal surface, but with no lashes at the margin ; (*c*) near the mesal angle, on the margin of either lid, a slight elevation, in which is the orifice of the LACHRYMAL CANAL.

   a. With the forceps grasp the skin at the mesal angle and pull laterad. With the scissors cut through into the orbit, keeping close to the bone. This will transect the two lachrymal canals.

   b. Under favorable conditions as to time, patience and skill the canals may be traced to the LACHRYMAL SACK, which receives also the canal from the other lid, and is continued into the cavity of the nose. Through this duct is carried into the nose any superfluous moisture on the surface of the eye ; the smallness of the orifices at the margins of the lids prevents the entrance of dust which might clog the passage.

§ 6. *The Third Eyelid.*—At the mesal or nasal angle of the eye between the lids, is a fold of mucosa, the PLICA or nictating membrane or third lid. Grasp the free margin and draw it laterad over the ball. The plica is easily seen in birds ; in man it is rudimentary and represented by a slight fold of mucosa.

§ 7. *Removal of the eye.*—Grasp either lid with the fingers or forceps and with the scissors cut about the ball close to the margin of the orbit. Cut deeper and deeper, dividing the fat and muscles, and lastly the firm, white, cylindrical OPTIC NERVE, at the bottom or apex of the orbit where it passes through a foramen in the cranium to join the chiasma ; Pl. XIX.

§ 8. *The Orbit.*—The margin of the ORBIT, the cavity containing the eyeball may be felt through the skin and its form and limits should be noted on the skull.

Compare the margin of the orbit with that of the prepared skull, and note that in the latter it is incomplete for a short distance at the lateral side, but that in the head under examination the gap is closed by a fibrous band which may be cut with the scissors ; Pl. XVI, *1, 2.* Ascertain that the mesal wall of the orbit, its roof and part of its floor are bony, but that the rest of the floor and its lateral wall are fleshy. In man and monkeys these parts of the wall are also bony, so that the orbit is completely circumscribed in the prepared skull.

§ 9. *The Muscles of the Lower Jaw.*—On Fig. 1 the word *cranium* corresponds nearly with a line across the side of the head about 5 mm. dorsad of the orbit and auditory meatus ; cut along this line, not too deeply at first, until the bone is reached. This is the CRANIUM, the bony case for the brain. The transected muscle is the TEMPORAL, one of the FLEXORS of the MANDIBLE, or lower jaw.

   a. Dissect it up from the bone, and note that it arises partly from the smooth, convex surface of the cranium, partly from ridges or crests of bone at the margins of the muscle.

   b. With young cats the crest at the dorsal margin is widely separate from its opposite, but with age they approach, especially caudad, and for a greater or less distance may unite to form a single, mesal crest. In some animals (*e. g.*, lion, hyena, gorilla) this mesal crest is considerably elevated, and the temporal muscles are correspondingly thick and powerful.

   c. In man the area covered by the temporal muscle is comparatively slight. Its extent and the action of the muscle may be felt if the fingers are pressed upon the temple while the mouth is opened and closed.

§ 10. A line from the middle of the eye to the auditory meatus corresponds nearly with the ZYGOMA or cheek bone ; it may be felt under the skin in ourselves and is prominent in certain races and emaciated persons. From it arises the MASSETER, the second great mandibular flexor. Determine its ventral border with the finger and tracer, and cut along it carefully so as not to injure the blade.

§ 11. Transect the temporal and masseter on the other side in like manner. Grasp the mandible and work it up and down ; then with the forceps pull upon the cut end of the part of either muscle which is attached to the mandible ; the movement of the latter illustrates the action of the muscles by contraction.

§ 12. *Functions of the Teeth.*—Work the mandible and note the relations of the mandibular and maxillary teeth. The MANDIBULAR CANINE (Pl. XVI, *C*) enters the DIASTEMA, the interval between the maxillary canine and the incisors ; the other mandibular teeth pass mesad of the maxillary ; the last (MOLAR) in the mandible (Pl. XVI, *M*) acts against the last premolar of the maxilla like a scissors-blade ; hence these two are called SECTORIAL TEETH. Their office is to cut the flesh ; the canines hold and lacerate the prey, while the incisors are mainly used in gnawing bones.

§ 13. *The Oral Cavity.*—Force the mouth open to its fullest width. Draw out the tongue and look down the throat ; there should be visible the free margin of the SOFT PALATE dorsad, and ventrad the triangular point of the EPIGLOTTIS. Defer the examination of other features till the mandible is removed.

a. Looking into one's own mouth before a mirror the soft palate will be seen to present a mesal appendage, the UVULA, which is absent in the cat.

§ 14. *The Arrangements of Parts in the Neck.*—See that the cut caudal surface of the neck is as smooth as possible. Compare with Plates XVI and XVII ; of course on the latter only the mesal parts are shown.

a. The muscles are easily recognized ; their thickness in the dorso-lateral region enables the cat to not only support its head in what would be an impossible attitude for us, *viz.*, with the neck erect while the body rests on one side, but also to carry prey and its young.

b. Make an outline diagram of the cut surface of the neck as follows : Represent the outline of the whole by an approximately circular line, twice the actual size. Indicate the MESON by a "meridian," a vertical or dorso-ventral line at the middle of the circle ; parts of this line may be erased when the drawing is finished.

c. At the proper place on the meson, commonly near its middle, represent the AXON, (body-axis) by a depressed circle ; if the transection passed between two vertebræ the surface will be rather soft and smooth ; if through the body or CENTRUM of a vertebra, then it will be hard and rough.

d. Dorsad of the axon is a subcircular space, the NEURAL (spinal) CAVITY ; the general outlines of the bony arch may be indicated.

e. The NEURAXIS is represented by the nearly cylindrical MYEL or spinal cord ; the exact form and the details of its structure will be studied later.

f. Near the ventral side, covered by thin muscles, is the TRACHEA or wind-pipe ; the cartilaginous rings of which it partly consists cause it to maintain its form. It is approximately circular, but since the dorsal side of each ring is membranous that side is slightly flattened.

g. But the walls of the ESOPHAGUS, just dorsad of the trachea, are composed of muscle and mucosa, and therefore collapse ; on the drawing it should be represented as a flattened or corrugated circle.

h. Laterad of the interval between the trachea and esophagus at either side are two very important structures, the CAROTID ARTERY and the VAGUS (pneumogastric) NERVE. The artery may contain some blood and if not may be recognized from its tubular shape, more or less collapsed.

i. The vagus of the cat, in the neck, is contained in the same sheath as the SYMPATHIC (sympathetic); in man the two are distinct.

j. The two jugular veins, ectal and ental, may commonly be recognized from containing some blood.

§ 15. *The Trachea.*—Dissect off the muscles covering the ventral aspect of the trachea ; cut off a piece of it, 1-2 cm. long and note that (*a*) its cartilaginous rings alternate with softer tissue, and (*b*) the ring are not complete, the interval at the dorsal side being occupied by muscle and fibrous tissue.

§ 16. *The Larynx.*—Cephalad the trachea enlarges and is modified to form the LARYNX, less prominent than in man where its projection is called the "Adam's apple." Just cephalad of the larynx is a slender bone, or rather chain of bony and cartilaginous segments, the HYOID BONE, Pl. I ; it is in the form of a U, its ends being attached to the base of the skull just laterad of the hemispherical bony elevation, the TYMPANIC (or auditory) BULLA (Pl. XVI).

§ 17. *Removal of the Neck.*—Consult Plates I and XVI. Flex the head and neck upon one another ; with the arthrotome divide the muscles just caudad of the lambdoidal crest ; this will permit further flexion ; continue to cut till there are reached two bony projections the OCCIPITAL CONDYLES by which the skull articulates with the ATLAS (first vertebra).

a. Work the parts on each other. Cut the capsule inclosing each joint ; then the membrane at the meson, which will expose the NEURAXIS at the junction of the myel with the brain. Divide it and pull away the vertebræ and attached muscles from the ventral soft parts of the neck, esophagus, trachea, etc.

§ 18. *Removal of the Mandible* (lower jaw).—Unless the student has done this before, the two halves (RAMI) should be removed separately.

a. With the scalpel cut along the mesal (inner) side of the bone to the SYMPHYSIS (mesal union between the tips of the rami). In young animals the symphysis may be divided with the arthrotome or pocket knife ; in older the coarse scissors or even nippers may be needed. In man the rami are closely united at an early period.

b. Work the ramus up and down and note that the attachment of the ARTHRAL CONDYLE (Pl. XVI) is inclosed by a FIBROUS CAPSULE. Push a narrow, strong blade directly mesad for 10-15 mm. and continue to cut cephalad and caudad till the bones are freed from one another. The ramus may then be twisted and pulled away. Repeat with the ramus of the other side.

§ 19. Pull the trachea and esophagus caudad and ventrad and dissect them away from the base of the skull as far as the meatus.

a. With the scissors slit the dorsum of the esophagus along the meson. It expands cephalad as the PHARYNX (Pl. XVII).

b. Pull the trachea caudad. Pass a probe cephalad and note its emergence into the pharynx through the GLOTTIS (the narrowed orifice of the larynx ; also, cephalad of the orifice, the triangular, cartilaginous EPIGLOTTIS, longer and more pointed than in man.

§ 20. *Removal of the Larynx and Trachea.*—Note the U-shaped bone at the side of the pharynx, the hyoid. Cut between it and the larynx, and remove the latter with the trachea (§ 15). Trim off the remnants of the pharynx and esophagus.

§ 21. *The Cartilages of the Larynx.*—Cut and tear off the small muscles upon the ventral aspect of the larynx and note that it consists mainly of two cartilages, the CRICOID caudad, resembling an enlarged tracheal ring, and the THYROID cephalad, larger and more irregular.

a. Work the thyroid to and fro and note that it has considerable mobility upon the cricoid.

b. Medisect the trachea and larynx with the scissors upon both the dorsal and ventral sides. Note that the cricoid is a complete ring, wider at the dorsal side. Also that upon its cephalic margin, at either side of the meson, is perched a small cartilage, the ARYTENOID.

·§ 22. *The Vocal Bands.*—From the arytenoid cartilage to the thyroid, near the root of the epiglottis, extends a fold or ridge or shelf of mucosa, the VOCAL BAND.

a. The common name, *vocal cord*, is misleading; a fold of mucosa is supported by fibrous tissue.

§ 23. *Action of the Vocal Bands.*—Hold either half of the trachea and cricoid firmly. Tilt the thyroid ventrad and note that the band is rendered more tense ; tilt the thyroid dorsad and the band is relaxed. Move the arytenoid from side to side and note, when the two are tilted mesad, that the vocal bands are correspondingly approximated.

a. With a fresh, entire specimen, if a tube is tied in the trachea and the larynx compressed somewhat so as to approximate the bands, blowing through the tube will produce a vocal sound on account of the vibration of the free margins of the bands.

b. If time permits make an enlarged drawing of the mesal aspect of the larynx ; compare Pl. XVII.

§ 24. *Removal of the Tongue.*—Pass a scissors-blade from the mouth caudad into the pharnyx and cut first on the right side and then on the left ; unless the hyoid bone has been previously dissected out it will be transected.

§ 25. *The Lingual Papillæ.*—At the tip and base they are simple and FILIFORM, short and closely set at the tip, longer and scattered at the base ; in Pl. XVII they are not named or very distinctly shown. On the longer intermediate region are the horny, sharp and recurved ODONTOID PAPILLÆ ; with a lens may be traced the transition between them and the filiform. At various points but especially near the middle of the tongue are the blunt FUNGIFORM PAPILLÆ. A little caudad of the middle are about half a dozen CIRCUMVALLATE PAPILLÆ, arranged as a V with its open end toward the tip ; each may be described as a fungiform papilla encompassed by a circular trench and wall. At the margin near the base are six or eight papillæ set like a fringe. The tonsils are not very distinct in the cat.

§ 26. *Transect the Tongue.*—Note its muscular mass, the mesal rhaphe (seam) and the thick dorsal mucosa.

§ 27. *The Palates.*—On the roof of the mouth note that the HARD PALATE, between the teeth, presents 6-8 transvers RUGÆ, ridges of papillæ, with other papillæ, less regularly placed, in the intervals. There is a mesal papilla of a different shape just caudad of the interval between the incisors. The SOFT PALATE is smooth.

§ 28. *The Eustachian Tube.*—With the scissors cut away the soft palate, exposing the postnasal cavity. At either side is an oblong orifice, straight or slightly crescentic, the orifice of the EUSTACHIAN TUBE leading to the tympanum or middle ear.

§ 29. *The Tympanodisk* (membrana tympani).—Trim the remnant of the external auditory meatus (Pl. II) close to the bone. Hold the head so that the light enters it and note that it is closed by a membrane, obliquely placed, and crossed by a light bar. The membranous septum is the TYMPANODISK, or MEMBRANA TYMPANI, and the bar is the handle of the MALLEUS, one of the three ossices of the ear.

§ 30. *The Tympanum* (middle ear or drum of the ear).—Ventrad of the meatus is a rounded elevation, the tympanic (or auditory) bulla (Pl. XVI). The ventro-caudal part is a thin shell of bone which may be opened with the nippers or a stout pocket-knife. The considerable cavity is lined by a delicate mucosa.

a. Remove the caudal wall as completely as possible ; look in at the caudal end and note a semilunar orifice leading cephalad. Just mesad of this orifice is a circular depression, the FENESTRA ROTUNDA.

b. Pass a probe very carefully into the semilunar orifice, looking at the same time into the meatus ; the probe will be seen through the transparent tympanodisk.

§ 31. *Opening the Tympanum Proper.* With nipper and coarse scissors cut away the thin septum between the two cavities ; avoid injuring the tympanodisk as long as possible, at any rate until there is recognized the attachment of the handle of the malleus to its ental surface. Then the margins of the cavity may be nipped away so as to expose it completely.

§ 32. *The Auditory Ossicles.*—The long handle of the malleus forms an angle with its head. Attached to an intervening neck is the short tendon of an almost spherical muscle, the TENSOR TYMPANI.

a. Move the handle of the malleus and note the communication of the movement to two other bones, the INCUS and STAPES and the attachment of the "foot" of the latter at the FENESTRA OVALIS.

b. Extract the ossicles and examine with the magnifier, if possible while resting on a dark surface.

§ 33. *The Eustachian Tube.*—From the pharyngeal orifice (§ 28) pass the tracer cautiously dorso-caudad toward the tympanum, keeping the concavity of the tracer ventrad ; the point will presently enter the tympanum at the side of a projecting shelf of bone.

§ 34. *The Semicircular Canals.*—These and the other parts of the labyrinth (ental or internal ear) are inclosed in dense bone. With fine nippers the bony tube containing one of the canals may be opened, but the parts are too small in the cat for examination in this connection ; see *Anatomical Technology*, 529-533, Fig. 127.

§ 35. *The Nasal Cavity.*—With the scalpel cut off the end of the nose close to the bone ; note the mesal cartilaginous NASAL SEPTUM ; in Pl. XVII the septum has been removed. At the sides of the septum are the convoluted cartilaginous continuations of the TURBINALS or turbinated bones.

a. Upon some of these are distributed the olfactory nerves ; they are more complex and abundant in the cat and especially in the dog than in man.

b. At either side of the septum introduce the probe and push it caudad, keeping close to the floor of the nasal cavity, and it will emerge in the pharynx on the same side.

§ 36. *The Lachrymal Duct.*—This has been mentioned in § 5 ; its nasal orifice is at a point ventrad of the *M* of the abbreviation *O. Mxtrb.* in Pl. XVII, but time and skill are required for tracing it.

Also some figure of a section of the eye-ball, such as contained in all works on Anatomy or Physiology ; a section of the cat's eye is represented in *Anatomical Technology*, Fig. 126.

§ 1. Review Pract. VI §§ 5–7, for the location of the eye, the form of the ORBIT, and the PLICA or third eyelid.

§ 2. The following directions and descriptions refer directly to the sheep's eye ; but that of the cat might be employed instead, and should be compared if possible.

a. If possible the eyes should retain the lids for a width of 1–2 cm., and special care should be taken to retain their mesal (inner or nasal) junction ; if the lids have been removed the directions in §§ 5–7 cannot be followed.

§ 3. *Determination of the Aspects.*—The cephalic (facial, "anterior," or ectal), whether or not partly covered by the lids, is smooth and more regularly convex and presents (in the sheep) an elliptical area surrounded by a brown line.

The caudal (cranial, "posterior," or ental) aspect may be hidden by masses of fat and by the muscles ; if these have been partly removed the remnants will still serve to distinguish this from the other.

§ 4. *The Eyelids.*—Note that their ectal surface is hairy, and that along the free margins are longer hairs, and less regular and less gracefully curved than the lashes of man ; they are more numerous on the upper lid, and from this may be determined the dorsal and ventral aspects of the entire organ. The angles of junction of the lids are the mesal and lateral CANTHI (commonly called "inner and outer").

a. The technical name for eyelashes is *cilia* (singular *cilium*); the same word is applied to the microscopical, structureless, moving filaments upon the mucosa of the air passages and some other parts.

§ 5. Between the eyeball and either lid insert a scalpel-handle, and note that its passage is checked at about 1 cm. from the margin. Insert a scissors-blade in the same way about one-third of the distance from either canthus and transect the lid ; repeat at one-third of the distance from the other canthus.

§ 6. *The Meibomian Glands.*—Reflect the middle third of the lid, demarcated as above and note that the ental surface, near the margin presents a series of dark stripes, 3–4 mm. long, corresponding with small orifices at the margin. These are the MEIBOMIAN GLANDS.

a. The Meibomian glands secrete an oily matter which anoints the margin of the lid and prevents the usually small amount of liquid between the eyeball and the lid from running over the edge upon the face. The action may be illustrated as follows :
Nearly fill two glasses with water. Wet the edge of one. Dry the edge of the other and anoint it with sweet oil or other oily or fatty substance. Then carefully pour in water till both glasses are full to the brim. The wetted brim permits the overflow at once but in the other glass the water may rise perceptibly above the rim before it passes over the oil.

§ 7. *The Conjunctiva.*—The smooth membrane lining the ental surface of the lids is the CONJUNCTIVA. Note that it is continuous with the ectal hairy skin at the smooth margin of the lid, just as the mucosa of the mouth is continuous with the skin at the lips.

a. Note also that it is reflected from the lid upon the surface of the ball. When the lids have not been retained with the eye the cut edge of the conjunctiva may be traced, and in places lifted slightly with the tracer or forceps.

§ 8. *Sensitiveness of the Conjunctiva.*—During life the conjunctiva is exquisitely sensitive to irritation by small particles like dust or cinders, though more tolerant of the contact of a larger surface like the finger-tip. Most operations on the eye are now rendered painless, so far as concerns the conjunctiva, by the application of a few drops of a solution of cocaine.

a When a cinder or other irritating particle lodges upon the eye, rubbing should be avoided. Hold the upper lid down with the finger-tips applied at its edge. If after a few minutes the irritation does not cease, hold the lids far apart and dash water upon the eye. If this fails to wash the particle remove it as follows : Provide a rounded point like that of a lead pencil that has been used a little. Before a mirror draw the lower lid down ; if no foreign body is visible grasp the edge of the upper lid firmly and turn it up, if necessary over a toothpick or pencil. When the particle is seen, touch it *lightly* with the rounded point above mentioned and it will usually adhere to it. Of course the operation is more easily performed by another person standing behind the seated patient, and cocaine may be used if the conjunctiva is already inflamed or the patient is very apprehensive ; but if cocaine were accessible a physician could probably be consulted. It is worth bearing in mind that if the irritating particle has been in the eye for some hours the inflammation may persist even after its removal, so that the light should be excluded by a bandage.

b. For the inflammation above mentioned, or for dryness or redness of the conjunctiva, a simple and harmless remedy is a solution of *boracic acid* in water, five grains to the half-ounce (table-spoonful); when dissolved a few drops may be introduced either with a dropper or with the tip of the finger applied at the mesal canthus, and repeated as frequently as desired.

§ 9. *The Plica.*—Transect the other lid at about its middle. When the two lids are separated as far as possible there will be seen at one canthus a semilunar fold, the PLICA (third eyelid or nictitating membrane). This has already been seen in Pract. VI, and is outlined in Pl. XVI.

a. The plica is at the mesal (nasal or "inner") canthus ; hence the mesal and lateral aspects of the sheep's eye may be determined from it.

b. The plica extends to about the middle of the "lower" lid, but not so far along the "upper ;" hence from it may be also determined the ventral and dorsal aspects of the eye.

c. If, while a cat is sleeping, the lower lid be gently drawn down, the plica may be seen partly covering the eye before it is withdrawn.

d. The human plica (*semilunaris*) is an insignificant fold, an example of vestigial organs.

§ 10. *The Lachrymal Gland.*—On the dorso-lateral aspect of the eye *i. e*, diagonally across from the plica, ectad of the conjunctiva, perhaps covered in part by the cut margin of the skin of the upper lid, look for a pale, lobulated mass much like the parotid gland (Pract. VI, § 3, Pl. XV). Some or even all of it may have been cut away. This is the LACHRYMAL GLAND ; its secretion, the tears, a thin, slightly saline liquid, is poured upon the surface of the eye through ducts that open at or near the line of reflection of the conjunctiva upon the ball.

§ 11. *The Lachrymal Canals.*—Transect the skin about 1 cm. from the mesal canthus. Examine the loose tissue just entad of the skin for a pair of holes about 3 mm. apart. Pass the probe into either and it will emerge at the margin of the lid about 3 mm. from the canthus. Repeat with the other hole and note its emergence upon the other lid. These are the CANALICULI or lachrymal canals.

a. Were the adjacent parts retained these two canals would be found to open into a sack, the LACHRYMAL SACK, continuous through the nasal duct into the cavity of the nose.

b. If the cat's head has been retained these parts may be traced, though with some difficulty on account of their small size. The ventral end of the nasal duct is just laterad of the ventral turbinal, maxillo-turbinal, indicated in Pl. XVII by the abbreviation *Mxtrb.*

c. Through the passages above described the natural moisture of the eye, keeping the apposed conjunctival services soft, is drained away into the nose and evaporated by the currents of air. Where there is an excess of lachrymal secretion, as from taking cold, from the odor of an onion, or from laughter or crying, part runs over the edges of the lids as tears, and part makes itself apparent in the nasal cavity.

d. The orifices of the canaliculi at the margins of the lids are narrow and valvular so that particles of any size do not readily enter.

e. The small curved blunt pointed knife called by anatomists *syringotome* (tube-opener) is used by surgeons in opening up the lachrymal canals, and is called by them *canaliculus knife.*

f. The lids may be cut away with the scissors.

§ 12. *The Cornea and Sclera.*—On the free surface of the ball the elliptical area includes the CORNEA, transparent during life but rendered opaque by alcohol ; the rest of the surface of the ball is constituted by the naturally white and opaqe SCLERA, commonly called *Sclerotic.*

a. In the natural attitude of the sheep's head the long axis of the cornea is nearly parallel with that of the brain but not with that of the head as a whole.

§ 13. *The Caudal Aspect.*—Note again the following features more in detail than above (§3).

a. FAT and CONNECTIVE TISSUE, in white, irregular masses ; they constitute a cushion for the ball ; during prolonged illness or fasting the fat wastes so that the eye literally becomes "sunken."

b. In the midst, the white, firm, fibrous OPTIC NERVE.

c. The cut ends of the MUSCLES.

§ 14. *The Rectus Muscles.*—With the fingers (using the knife-point or tracer sparingly), pulling mostly from the center, tear out the fat and connective tissue so as to separate four muscles, at the four opposite sides of the ball ; these are the RECTI (straight) muscles. Their TENDONS unite to form a continuous sheet, thin but tough, entad of the severed conjunctiva.

a. There are two OBLIQUE muscles whose location and attachments are less easily recognized ; see the models and diagrams.

b. The superior oblique muscle passes through a fibrous loop at the mesal side of the orbit as through a pulley, and may be exposed if time permits.

c. Immediately surrounding the optic nerve is another layer of muscle, but it does not exist in man and may be disregarded.

§ 15. *Demonstration of the Actions of the Rectus Muscles.*—Hold the eye lightly between the left index and pollex, the dorsal side up. Grasp the rectus that is attached to the mesal side of the ball, *i. e.*, corresponding with the end of the cornea which is partly covered by the plica. Pull the muscle and thus rotate the ball mesad, *i. e.*, inward. Pull the muscle at the opposite side and rotate it laterad, *i. e.*, outward

a. Strabismus or "squint" may be due to either the shortness or undue contraction of one muscle, or to the length or weakness of its antagonist.

b. With the scissors trim off the rectus muscle and the plica.

§ 16. *The Optic Nerve.*—With the fingers and forceps tear apart the muscular masses surrounding the optic nerve and remove them with the scissors.   Note the fibrous constitution of the nerve and the firmness of the sheath.

a. Although the adult optic nerve is solid, excepting a small artery which traverses it, it was developed as a tubular outgrowth from the brain, as described in the lectures.

b. The place of attachment of the **nerve** is *eccentric, i. e.,* neare one side of the ball ; in the sheep it is ventrad of the center ; in man, mesad.

§ 17. *The Aqueous Humor.*—Compress the scleral part gently but firmly so as to render the cornea tense.   With a *very sharp* scalpel-point, (borrowed from the instructor if necessary) slit the cornea for the middle third of its long axis, not letting the point enter more than two mm , and relaxing the pressure just as soon as any liquid escapes ; this liquid is the AQUEOUS HUMOR ; as its name implies it is naturally watery and clear, but now it probably contains black particles dislodged by manipulation.

§ 18. *The Iris and Pupil.*—With the forceps carefully raise either cut edge of the cornea, making sure that only it is grasped, and with the scissors clip away the cornea, piecemeal, to within not less than 1 mm. of the brown boundary line.   This will expose the IRIS, a dark lamina, co-extensive with the cornea, and presenting a central orifice, the PUPIL.

a. In the sheep the pupil, like the cornea, is elliptical, and the long axis horizontal ; in the cat the long axis is vertical ; in man the pupil is circular.   The human iris varies in color, whence the name, signifying a rainbow.

b. The contraction of the pupil in response to an increase of light is familiar to all ; it may be observed by closing the eyes in a dark room, keeping them closed while entering a light one and approaching a mirror; when opened the dilated pupil rapidly diminishes in size.

c. When a cat is about to spring, even in play, the pupils commonly dilate.

d. Make a diagram of the cephalic aspect of the ball including the appearances now presented.

§ 19. *The "Anterior" Chamber.*—By inspection and careful use of the probe or tracer determine that the periphery of the iris is attached at or near the corneal margin ; the space between the iris and the cornea was filled, naturally, with the aqueous humor already mentioned, and is called the "anterior chamber."   The manipulation (§ 18) may have crowded the iris cephalad so as to nearly obliterate the interval.

a. Through the pupil will be seen part of the CRYSTALLINE LENS, naturally transparent but rendered opaque by the alcohol ; it will be examined presently.

§ 20. *Transecting the Eye.*—The other eye is to be transected like an orange, as follows : Hold firmly but with the least possible pressure ; at any point on what might be called the equator, apply the sharpest attainable blade and use with sawing movement till the sclera is divided and a drop of liquid escapes.   Then insert a scissors-blade not more than 5 mm.; cut for this distance, withdraw the blade and reinsert in the same way, until the first incision is reached.   As the two halves separate keep the cephalic side down and lift off the caudal half like a hemispheric lid or cup ; place the cephalic half carefully in weak alcohol.

§ 21. Examine the cephalic (ental) aspect of the caudal half. Near, but not at, the center is a spot of more or less distinct radiation corresponding with the place of attachment of the optic nerve (§ 16); in the drawing this should be below, indicating that it is toward the ventral side of the eye.

§ 22. *The Tunics or Coats of the Eye.*—If the cut edge is gently manipulated at any point it will separate into an ectal, white portion, the SCLERA (§ 12), and an ental, the CHOROID and RETINA.

a. The thick and fibrous sclera, with the cornea which is really only a transparent portion of the same tunic, gives firmness to the eye, protects the more delicate other tunics, and gives attachment to the muscles. In birds and in the sword-fish it is more or less completely ossified.

§ 23. *The Retina.*—Dip the specimen in alcohol. With the tracer, applied near the ventral cut edge, scrape the ental surface gently. There should be separable a delicate, almost filmy layer, the RETINA, constituting the ental of the three tunics.

§ 24. *The Choroid.*—Between the retina and the sclera is the middle tunic, the CHOROID, black and firmer than the retina.

§ 25. The relative positions and general characters of the three tunics are easily recognized, but some confusion may arise (and indeed existed in former issues of these directions) from conditions now to be stated.

a. Part of the retina really consists of two layers, an ental white and soft, and an ectal, jet black in color from the abundant pigment.

b. Even in those regions where the pigmentary layer exists parts of the latter sometimes separate from the white layer and adhere to the choroid, appearing to constitute a part of it.

c. The ental surface of the sclera is pigmented and brownish (*lamina fusca*).

§ 26. If care be taken the retina can be detached from the choroid at most points, and drawn or pushed toward the center. .

a. Note the absence of the YELLOW SPOT which, in man, occupies the center just opposite the pupil.

b. Ventrad of the center (mesad in man) is the place of entrance of the optic nerve, of which the retina is really the expansion, constituting the BLIND SPOT.

§ 27. *The Tapetum and Fovea.*—When the retina has been removed, part of the choroid will be seen to present a rich purple color, the TAPETUM.

a. The tapetum is absent in man, but present in many animals. The corresponding area of the retina lacks the pigmentary layer (? 25) so that in the living or fresh eye, the tapetum shows through it and reflects even a very faint light; hence the glare that in the darkness is so startling in cats and many other animals. The common impression that this "eye-shine" is peculiar to dangerous species has sometimes caused needless terror to persons or led to the killing of innocent quadrupeds.

b. In man, on the other hand, at the middle of the retina, just opposite the pupil, is a depression which, from its color, is called the VELLOW SPOT, and which is the place of most distinct vision; in the sheep there may be a slight depression, the FOVEA, but not the yellow color.

§ 28. *The Vitreum.*—In the cephalic half of the alcoholic eye is a jelly-like mass; this is the VITREUM or vitreous humor; it is naturally transparent and nearly liquid, much like the white of egg before boiling. It is inclosed in a delicate membrane or capsule. Holding the specimen tilted a little, with the tracer push the vitreum at one side gently toward the middle. It will separate readily from the retina which lines the caudal portion of this half of the eye, but somwhat less so from a dark circular strip of the choroid presenting numerous radiating furrows and ridges, the CILIARY PROCESSES.

§ 29. *The Lens.*—With the scissors cut the capsule of the vitreum and note, imbedded in the mass, the CRYSTALLINE LENS, already seen through the pupil (§ 19). Continuing to use the scissors very carefully detach the entire vitreum from the lens ; the capsule of the former is so closely connected with that of the latter that there will be danger of displacing the lens.

a. The lens is naturally clear like glass. The condition called *cataract* is due to opacity such as in the preserved eye is caused by the alcohol.

. b. If the lens is torn apart with the nails it will be found to separate into concentric layers somewhat like those of an onion ; the central portion may be still transparent.

§ 30. The CAPSULE OF THE LENS is really very firm ; the two layers from the cephalic and caudal faces unite near the margin to form the SUSPENSORY LIGAMENT ; this is attached at its periphery, and is relaxed by the contraction of the ciliary muscle so as to permit the lens to become more convex ; this is explained in the lectures under the head of Accommodation. The ligament is well shown in a preparation (No. 2969) of the dog's eye, made by Dr. Fish.

a. After the removal of the lens a drawing should be made.

§ 31. The natural conditions of the transparent mediums of the eye can hardly be appreciated from alcoholic specimens, and fresh eyes should be examined. Sections should also be made in the other two planes, *i. e.*, medisections and longisections.

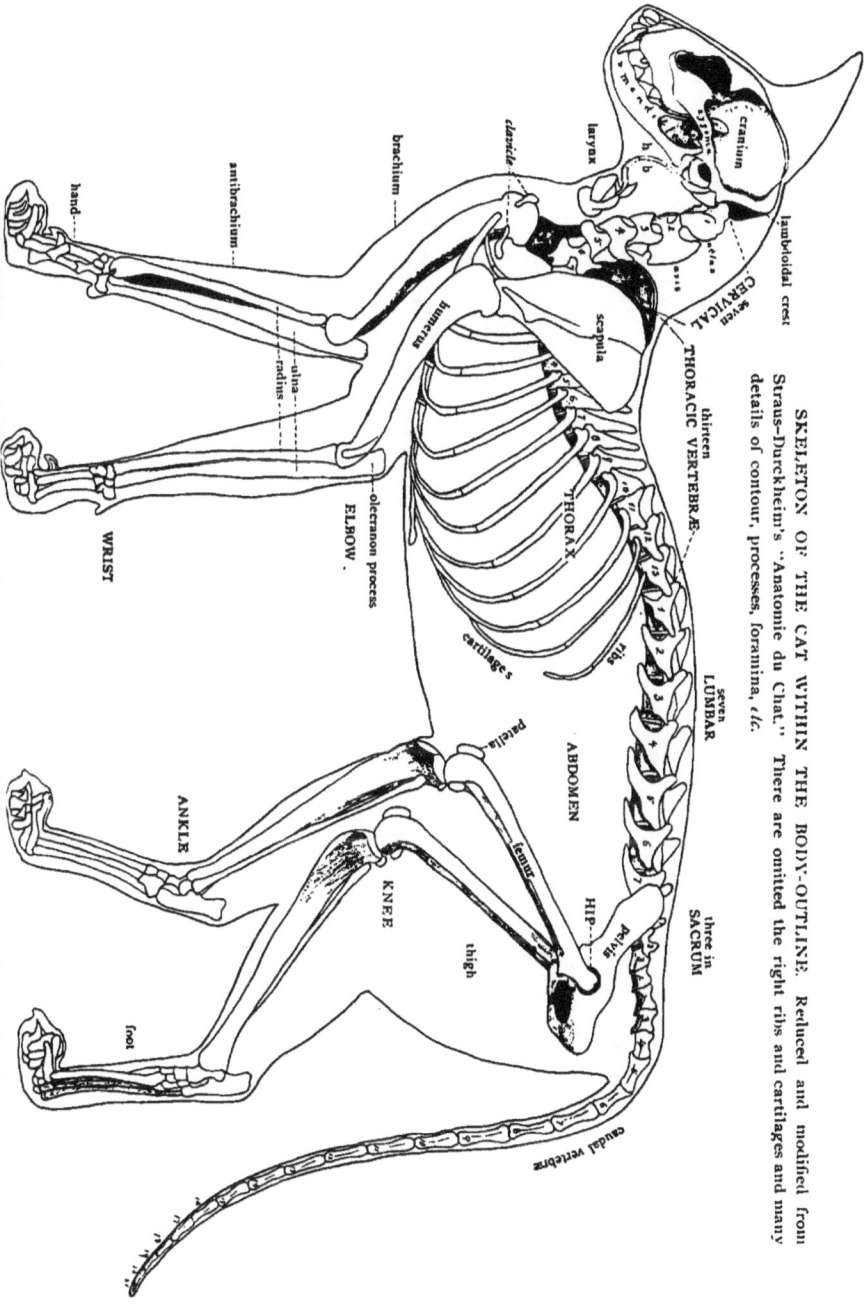

SKELETON OF THE CAT WITHIN THE BODY-OUTLINE. Reduced and modified from Straus-Durckheim's "Anatomie du Chat." There are omitted the right ribs and cartilages and many details of contour, processes, foramina, etc.

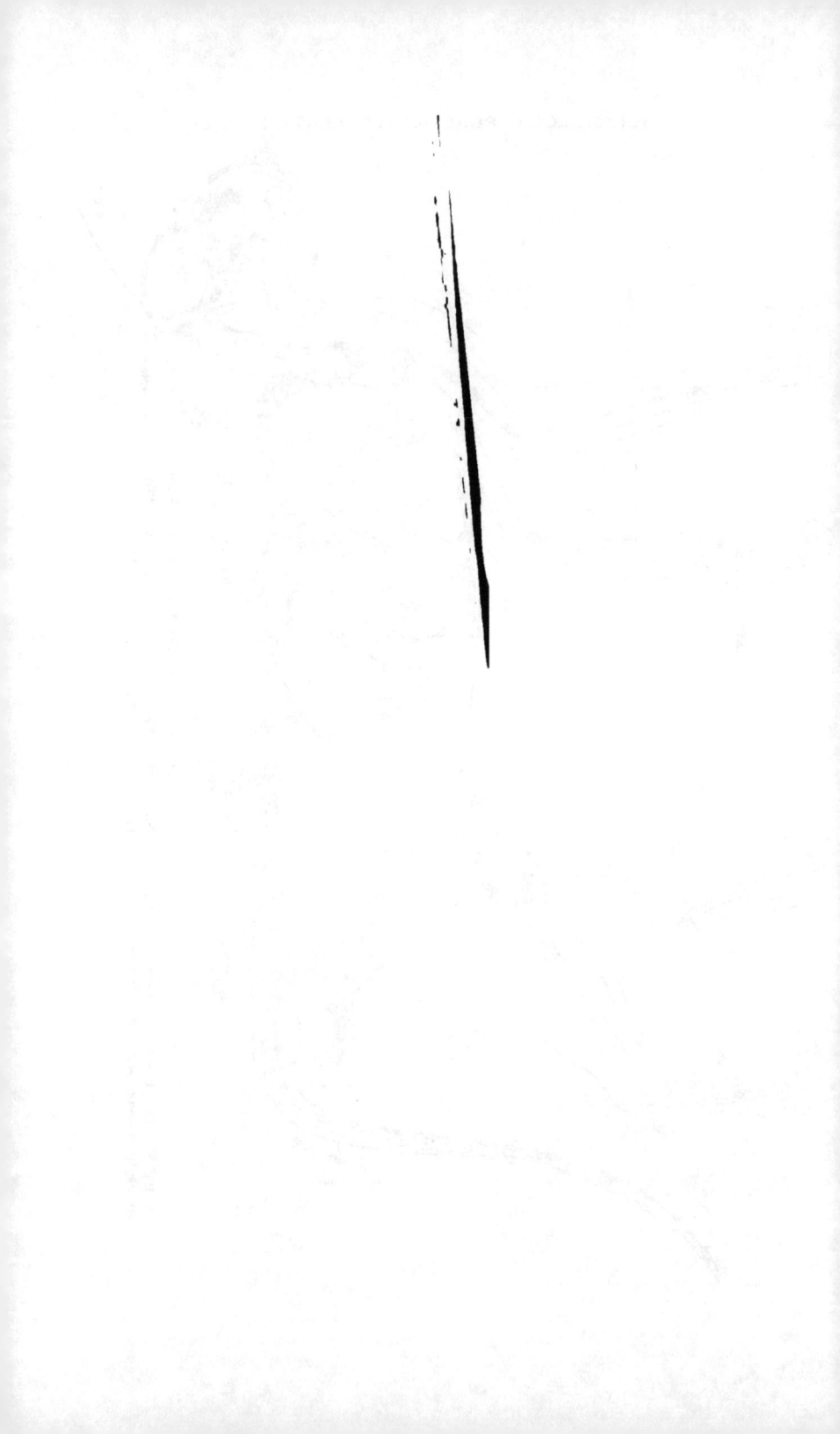

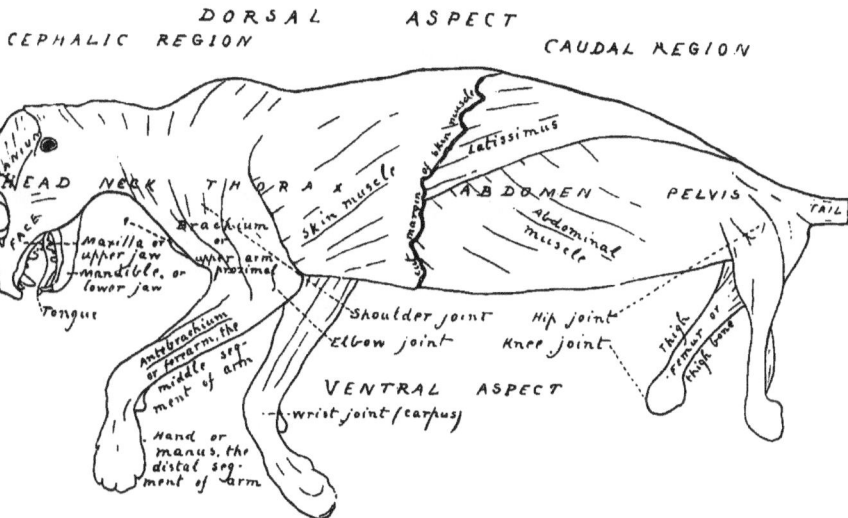

THE LEFT SIDE OF THE CAT less than one-half natural size   The TAIL was cut short   The SKIN has been removed, together with the caudal ("hinder") portion of the SKIN MUSCLE.   The cut margin of the skin muscle is indicated by the double line across the body.   The ARMS (front legs) are entire ; the LEGS (hind legs) were amputated at the KNEES and most of the flesh removed from the THIGHS   The left EYE is recognizable just in front of the word FACE.   The EARS were removed with the skin, but the dark spot above the word HEAD represents the tube or EXTERNAL AUDITORY MEATUS

1, the cephalic end of the STERNUM (breast-bone) covered by muscle.

Defects.—At the tip of the MAXILLA (upper jaw). the left CANINE TOOTH should be pointed like the right, and should be demarcated from the jaw by two lines indicating the margins of the gums.

The word *antebrachium* should be *antibrachium*.

The BRACHIUM and ANTIBRACHIUM are proximal and distal segments of the arm proper or commonly so-called ; but see the qualification in the explanation of Figure 2. Lecture I.

VENTRAL ASPECT OF THE THORACIC REGION OF THE CAT, reduced. The animal is supposed to be on its back with the arms outstretched ; the direction of the arms is such as might be assumed by the animal in climbing a thick tree ; the shoulders are widened so as to resemble more nearly those of man ; the muscles are rendered tense so as to be more readily distinguished and divided.

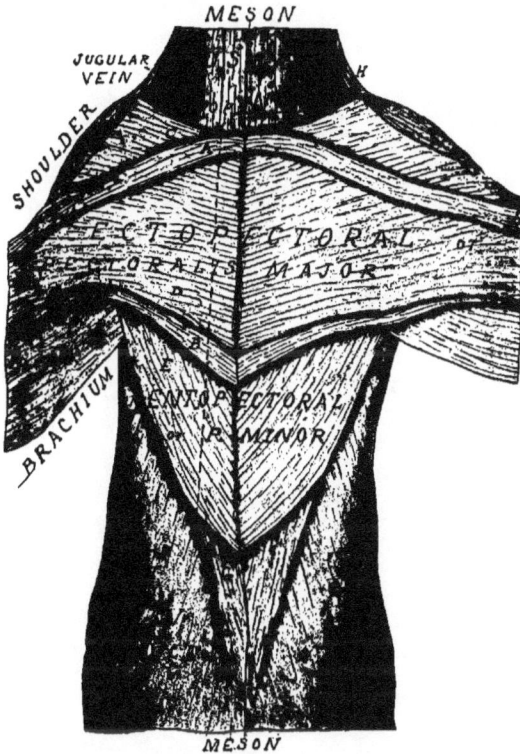

The muscles of the neck, abdomen and brachium are vaguely indicated. Most of the muscles here shown constitute the group called PECTORALS. The pectorals form two layers, an ectal (superficial), the ECTOPECTORAL ; an ental (deep), the ENTOPECTOR-AL ; with the former the general direction of the fibers is approximately transverse ; with the latter, obliquely latero-cephalad. In man, the raccoon and a few others, the ectopectoral is much the larger, so that the names commonly applied (*pectoralis major and p. minor*) are appropriate ; but in the cat, as in most mammals, the reverse is the case, and only the more cephalic portion of the entopectoral is covered by the ectopectoral

Each of the pectoral muscles in the cat consists of several laminæ more or less easily separable ; besides those shown in the figure (A, B, C, D, E, G,) there is a seventh, a part of the entopectoral, which is entirely hidden.

H indicates a ribbon-like muscle which overlaps the pectorals at the shoulder and on the brachium ; at about the middle of its length lies the CLAVICLE (collar-bone, Plates I and II), small in the cat ; but in man it is large and the cervical and brachial parts are separated by it.

The interrupted line X - - - X at the right of the MESON (middle line) indicates the incision to be made through the pectoral mass.

Defects.—The fasciculus marked A lies too far cephalad. On the neck the word *MUSCES* should be *MUSCLES*.

RIGHT AXILLARY REGION OF THE CAT AFTER DIVISION OF THE PECTORAL MUSCLES.

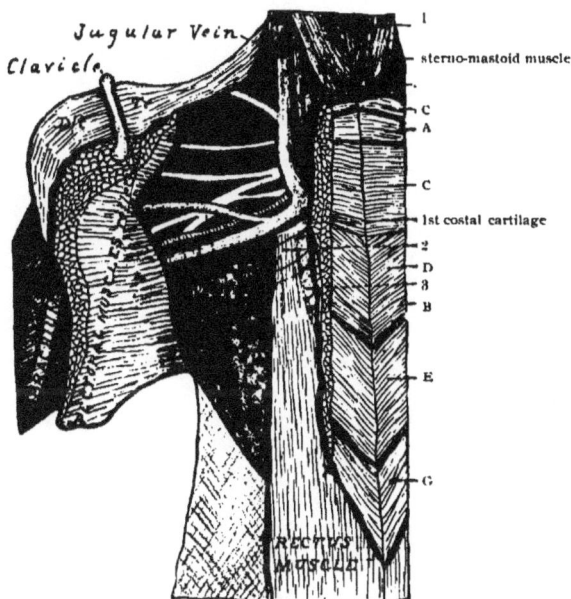

The PECTORALS have been transected along the line X X as indicated in Pl. III. The distal portions have been reflected laterad upon the shoulder. The muscle marked *H* in Pl. III is everted so as to expose its ental surface and the CLAVICLE (collar-bone) attached thereto ; the name points to the sternal end of the clavicle. The abbreviations *Dlt.* and *Tr.* are upon the two portions of the muscle, corresponding to parts of the deltoid and trapezius of man.

The main object of the figure is to facilitate the recognition of the great vessels and nerves which traverse the axillary space from the root of the neck to the arm. Farthest caudad is the AXILLARY VEIN, joined by a branch, and itself uniting with the ECTO-JUGULAR (external jugular) to form the BRACHIOCEPHALIC ; the unseen union of this with its opposite forms the PRECAVA seen in Pl. VII. Just cephalad of the vein is the AXILLARY ARTERY. A few nerve trunks are shown ; their actual number is greater, and their relations very complex, as may be seen from *Anatomical Technology*, Figs. 101, 102, 105, 106. The fat, connective tissue and smaller vessels and nerves are not shown.

The capitals A—G indicate portions of the pectoral mass similarly lettered in Pl. III.

1, a cervical muscle. 2, the muscular attachment of the RECTUS MUSCLE. 3, its thin tendon covering the second intercostals.

RIGHT ARM OF THE CAT FROM THE ULNAR (caudal or "inner") ASPECT. Some of the other muscles have been removed so as to expose the BICEPS. With the left arm the directions of parts are reversed.

1, cut surface of the muscle connecting the SCAPULA with the thorax ; 2, cut surface of muscle removed to expose the head of the HUMERUS : 3, LIGAMENT which crossed the groove (4) in which plays the TENDON of origin of the biceps ; the ligament has been divided and reflected ; 8, a small division of the TRICEPS, the great extensor of the antibrachium ; the rest of the triceps has been removed ; 9, tendon of insertion and distal portion of the BRACHIALIS ; 10, end of the OLECRANON PROCESS, the "point of the elbow ;" 11 and 12, cut surfaces of muscles ; 13, pad covering the PISIFORM BONE.

The biceps is selected as a nearly typical muscle, consisting of a fusiform, fleshy body or belly and two tendons, the proximal, of *origin*, the distal, of *insertion*.

The name *biceps* (two-headed) refers to its condition in man where there are two tendons of origin, one, the "long" or *glenoid*, from the margin of the glenoid cavity of the scapula, the shallow socket for the head of the humerus ; the other, the "short" or *coracoid*, from the tip of the coracoid process of the scapula. With the cat only the former, the "long" head, is present, but the name is retained.

The biceps is inserted upon the RADIUS near the elbow ; in this figure the

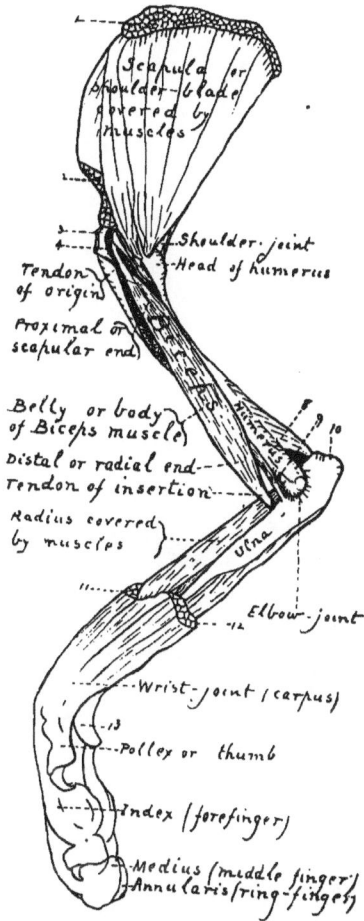

point of insertion is hidden by the ulna. The brachialis (9) is inserted upon the ulnar

Compare with the right arm as shown in Plates I and II and with the human arm. Note that the hand is not only flexed (bent) somewhat at the wrist, but that the palm looks in the same direction as the elbow points ; this is the condition when we place the hand, palm downward on the knee, or when we get on "all fours" and is technically called PRONATION ; it is the usual condition with quadrupeds. The cat and some others can rotate the parts somewhat into the condition of SUPINATION ; we can do this freely, completely and forcibly, as in turning a gimlet, cork-screw or screw-driver. In walking, the human hand is commonly *semi-pronated*, the pollex (thumb) forward.

THORAX OF THE CAT OPENED ON THE RIGHT SIDE.

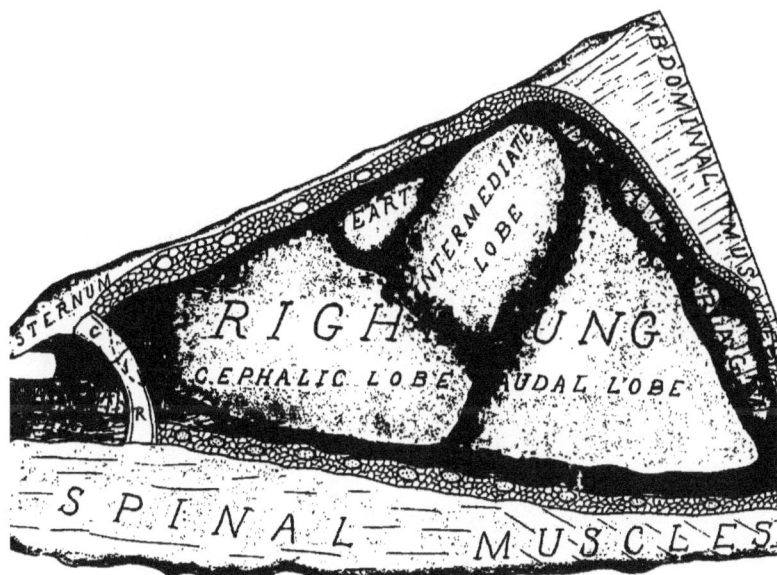

The parietes were removed by incisions along the interrupted lines shown on Fig. 2.

The first RIB and its CARTILAGE are entire ; the curved, interrupted line upon them indicates approximately the outline of the CEPHALIC LOBE of the LUNG. The other ribs and cartilages are transected ; the cut ends of the latter are left blank, of the former dotted.

The LUNGS were filled with alcohol and have therefore nearly their natural size. The ventricular portion of the HEART is seen covered by the PERICARDIUM and partly overlapped by the lungs.

THE CAT'S THORAX AFTER THE
REMOVAL OF THE RIGHT LUNG.

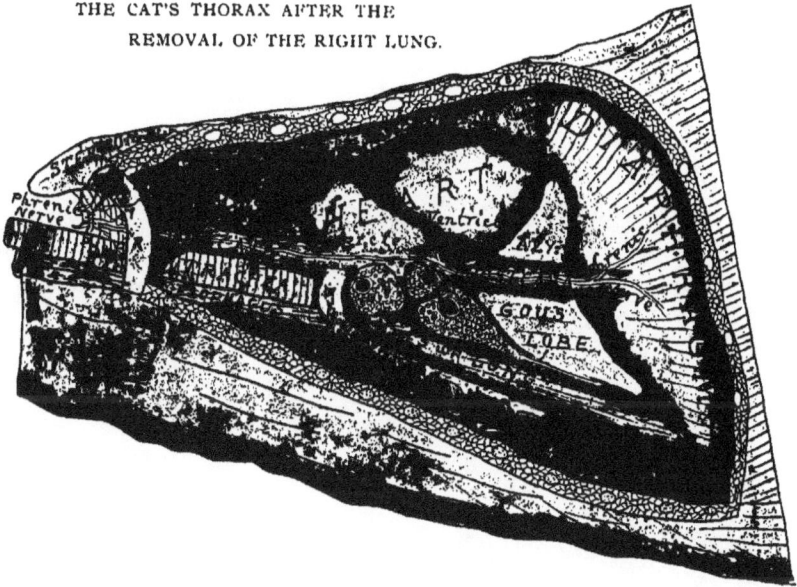

The viscera are undisturbed excepting that the three principal lobes of the right lung have been amputated. The large black spot on each root-section represents the BRON-CHIOLUS; the smaller the lesser AIR-TUBES, and the branches of the PULMONARY ARTERY and VEIN, without distinction. The heavy lines surrounding each root-section represent the PLEURA; the light extension from the third or caudal root is an exaggerated indication of the fact that here the two layers of pleura are unseparated by lung substance and constitute a MESOPNEUMON (Mpn). The fourth or AZYGOUS LOBE lies in a recess, partly covered by the POSTCAVA; the part ventrad of (*above* in the figure) the postcava is shaded a little darker to indicate that it is seen through not only its own layer of pleura but also the two layers forming the lateral wall of the recess. The extent of this recess is indicated upon the diaphragm, Plate VIII. The THYMUS BODY is shown large, as it is in young cats. The interval between the STERNUM and the thymus and HEART is shaded to represent two conjoined layers of pleura constituting the THORACIC SEPTUM. These layers diverge to extend either way upon the ventral and lateral walls of the thorax as the PARIETAL PLEURA; the heart and the thymus are also between two layers, but the outlines of these and indeed of all the other organs have been so sharply defined in the figure that the continuity of the visceral pleura over them is not well illustrated.

The heart is still covered by the PERICARDIUM, but the VENTRICULAR and AURICULAR portions are distinguishable; the latter joined by three VEINS, the POST-CAVA, PRECAVA and RIGHT AZYGOUS. (*On the figure this last word is misspelled azyous*). Upon the two cavas and intervening auricle lies the right PHRENIC, the motor nerve of the diaphragm, seen also cephalad of the first rib.

DIAPHRAGM OF THE CAT. CEPHALIC OR THORACIC ASPECT.

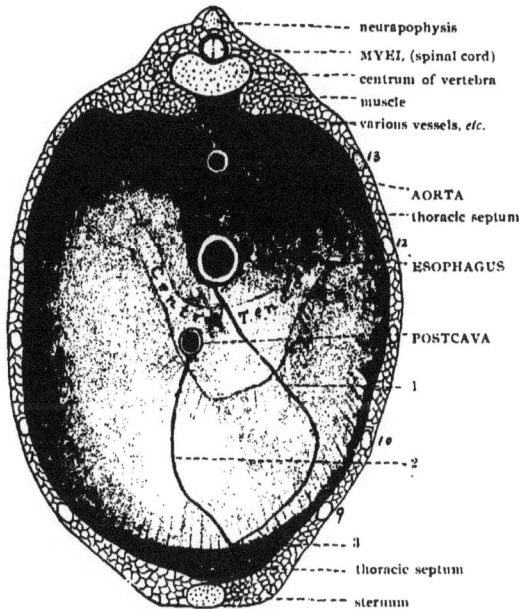

This is the view presented after the thorax has been cut away as at the close of Practicum III. The diaphragm is seen from the right side in Figs. 7 and 8 and the caudal (abdominal) aspect is represented in *Anatomical Technology*, Fig. 90 ; in that figure the dorsal side is down, here it is above. The ventral portion of the THORACIC SEPTUM appears in Pl. VII.

1, conjoined PLEURAS of right and left sides forming the left wall of the pocket for the azygous lobe of the lung seen in Practicum III, § 25.

2, right wall of the same ; this is attached to the postcava, and the interval between the postcava and the esophagus permits the connection of the azygous lobe with the rest of the lung.

3, interval between the thoracic wall and the ventral convexity of the diaphragm.

9, 10, 12, ends of the corresponding CARTILAGES ; the eleventh is crossed by the line from the postcava.

13, cut end of the thirteenth RIB.

Points illustrated.—A. The diaphragm is a dome, mostly of muscular fibers converging from the peripheral attachment to a CENTRAL TENDON.

B. It is traversed by three large tubes, the AORTA, ESOPHAGUS, and POSTCAVA.

C. The PLEURA (thoracic serosa) which covers its surface is reflected upon those tubes so that there is no crevice between them and the diaphragm.

D. The right and left sides of the thorax are separated by these tubes and by intervening double layers of pleura.

E. The general arrangement of organs and cavities which characterizes the vertebrates ; there is a dorsal cavity containing the myel representing the NEURAXIS (cerebrospinal axis) and a ventral containing the esophagus representing the ENTERON (alimentary canal) and chief blood-vessels

ABDOMINAL VISCERA OF THE CAT, EXPOSED FROM THE LEFT SIDE.

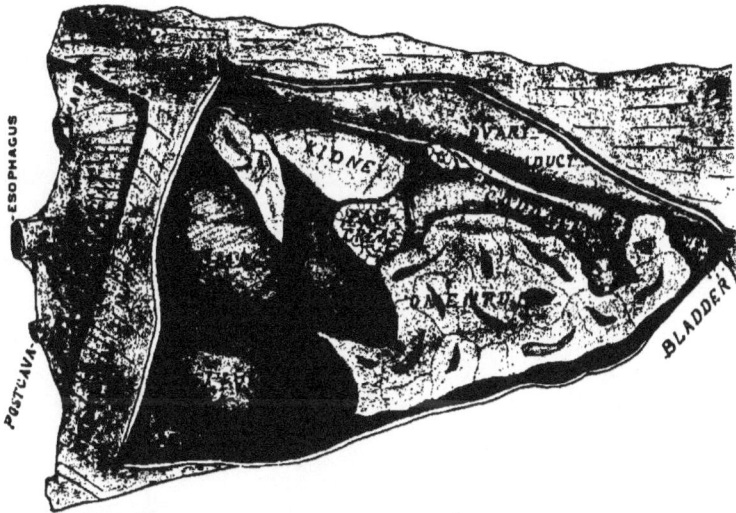

The dorsum is above ; the PELVIS and THIGHS are omitted. At the left (cephalad) projects the DIAPHRAGM with the stumps of the three traversing tubes already examined in connection with Pl. VIII.

The left wall of the abdomen has been removed and the dorsal edge everted. The viscera are undisturbed, but it must be borne in mind that the details of such a view of the more or less movable parts vary considerably in different individuals.

The ental surface of the parietes is formed by the smooth PERITONEUM. As will be seen during the dissection this is reflected at certain places upon the viscera so that, strictly speaking, all the organs are seen through it. The omentum is a fold of peritoneum, supporting fat and vessels.

Most of the parts are named. 1, a lobe of the liver, similarly numbered in Fig. 18 ; 2, part of the OMENTUM near the stomach ; 3, a fold extending cephalad from the OVARY ; 4, 5 coils of intestine.

The forms and connections of the organs will be seen in connection with Plates X and XI.

Defects.—The *h* of *stomach* and the *n* of *spleen* are obscured by too heavy shading.

ABDOMINAL VISCERA OF THE CAT AFTER REMOVAL OF MOST OF THE
SMALL INTESTINE.

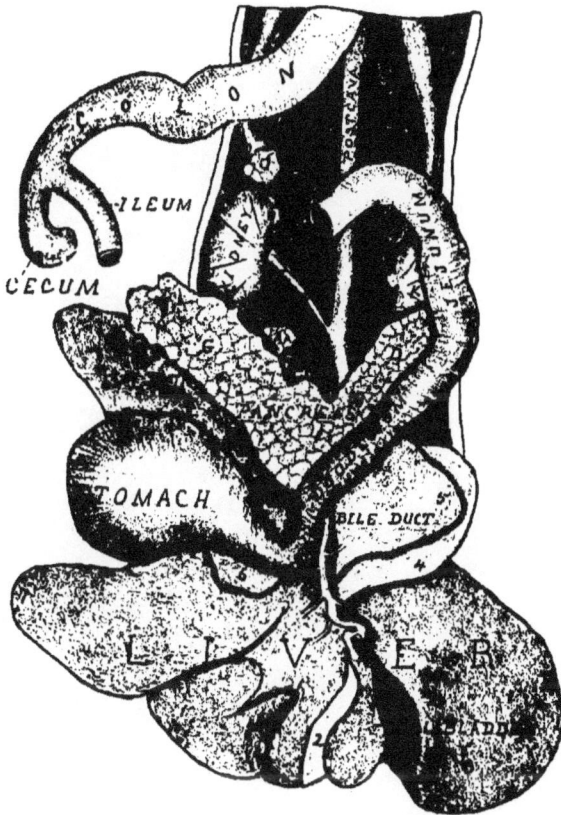

The specimen rests upon the back, *the cephalic end toward the observer.*

The lobes of the LIVER (1—7) are turned cephalad and spread apart ; the STOMACH
is rotated so as to conceal the esophagus ; the BILE–DUCT joins the DUODENUM ; the
pancreatic ducts are not shown. The COLON (large intestine) is turned to the left so as to
display the the CECUM : this is often shorter than here shown.

C, P, the CARDIAC and PYLORIC regions of the STOMACH.

G, D the GASTRIC (or gastro-splenic) and DUODENAL portions of the PANCREAS ;
the latter is less developed in man.

A, the left ADRENAL (supra-renal capsule).

K, the right KIDNEY, concealed partly by the pancreas and duodenum.

O, the left OVARY.

The mesentery, aorta, most of the renal vessels and the ureters are not shown.

PELVIC VISCERA OF THE FEMALE CAT.

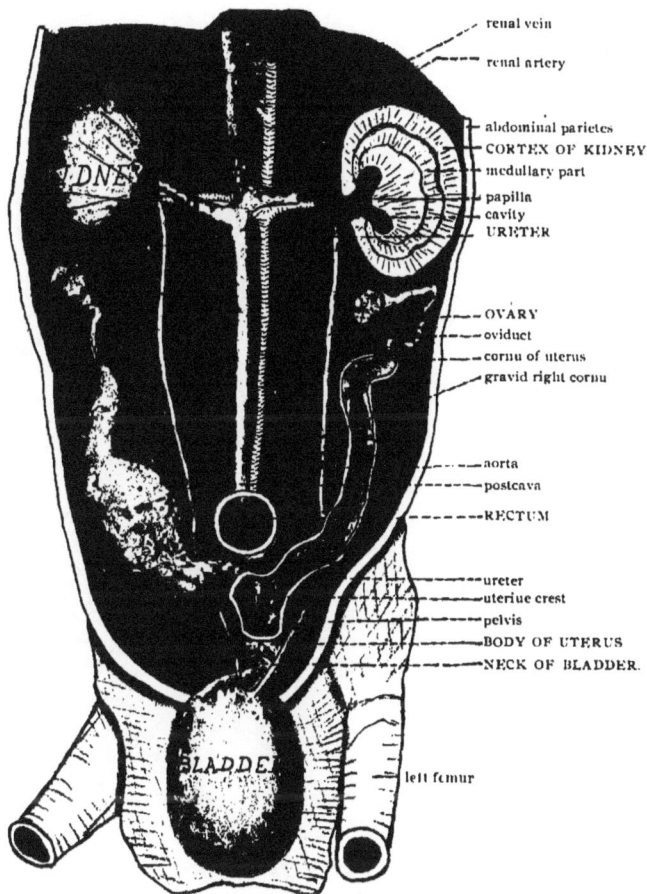

The COLON has been cut off where the RECTUM enters the narrow cavity of the TRUE PELVIS. The moderately distended BLADDER is turned caudad so as to show its narrow NECK, continued as the URETHRA ; also the attachments of the two URETERS. The left KIDNEY has been sliced off to expose the CORTICAL and MEDULLARY portions of the CAVITY, continuous with the ureter. The left OVIDUCT is uncoiled and the UTERINE CORNU opened. The right cornu is enlarged at one place as if containing an EMBRYO. The cut edges of the mesentery and some other details are disregarded. The right kidney is usually farther cephalad. The shading is too heavy.

THE SHEEP'S HEART.

VENTRUM.

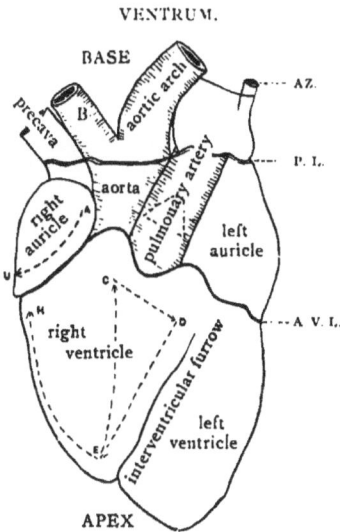

BASE

precava

B

aortic arch

aorta

right auricle

pulmonary artery

left auricle

AZ.

P. L.

right ventricle

interventricular furrow

A. V. L.

left ventricle

APEX

DORSUM.

Ventrum.—The upper figure is an outline diagram of the ventral ("front") and more familiar aspect of the heart after the removal of the pericardium, the attachment of which is indicated by the double line. The arteries (aorta, its main branch, B, and pulmonary artery) are distinguished by the transverse lines at the margins. On the right auricle the curved interrupted line A-U indicates the first incision for opening the cavity. On the right ventricle the lines connecting C D and E indicate the incisions for removing a triangular area without cutting the moderator band ; the line E H enables the intervening flap to be raised. On the pulmonary artery the broken line bounds the area to be cut out in order to show the valves.

Az. the Azygous vein. P. L., the pericardial line. A. V. L., the auriculo-ventricular line.

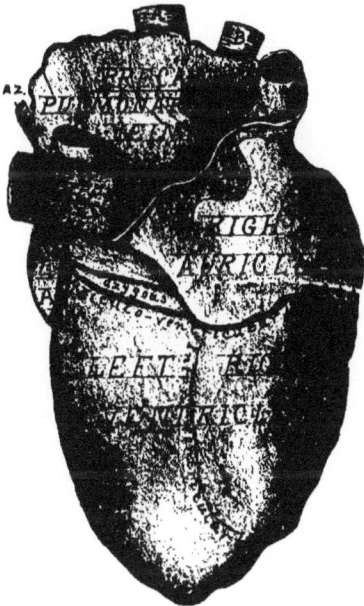

Dorsum.—This, the dorsal aspect, is much too heavily shaded. The word *azygous* is placed on what is really the CORONARY SINUS. The words PRECAVA and PULMONARY VEIN are written on a mass of fat and lung tissue remaining attached to the preparation. At the left *Az.* indicates the cut end of the (left) AZYGOUS VEIN. *L. A.*, part of the LEFT AURICLE. *A,* AORTA. *B,* its first great branch.

The RIGHT AURICLE should be crossed obliquely by a furrow, from the root of the azygous vein through the R of AURICLE and the G of RIGHT to the emargination at the root of the precava ; this emargination should be more distinct, as a notch. The furrow is the SULCUS TERMINALIS, demarcating the ATRIUM, into which the veins empty, from the APPENDIX. The broken white line is continuation of the line A U. on the Ventrum.

HEART OF THE SHEEP. THE RIGHT AURICLE OPENED ; ⨯ .8.

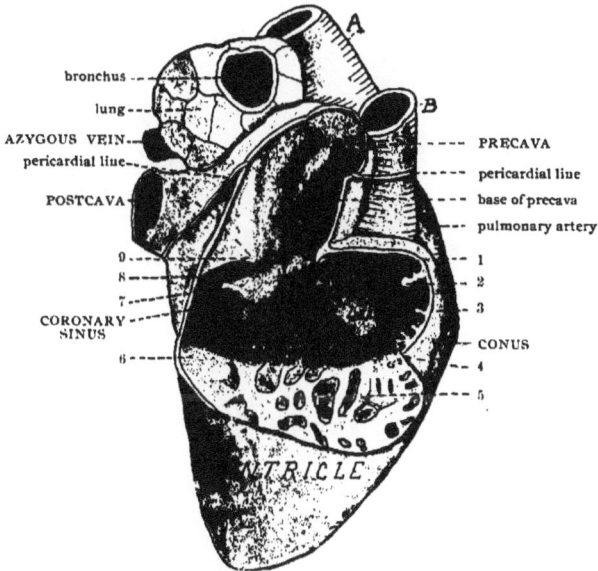

The preparation (2785) is viewed from the right side and obliquely, the apex away, so that the INTERVENTRICULAR FURROW, Pl. XII, does not appear.

The right wall of the PRECAVA has been wholly removed, but that of the AURICLE is turned caudad upon the VENTRICLE so as to expose the TRABECULÆ and intervening SINUSES which characterize the ventral or APPENDICAL part.

The AORTA (A) and its principal branch (B) have appeared in Pl. XII and from different aspects. The vessel marked PULMONARY ARTERY looks at first as if it were continuous with the aorta. The CONUS is the part of the right ventricle from which the PULMONARY ARTERY directly arises.

At the left of the figure the interrupted line from the upper (cephalic) margin of the postcava indicates its course into the atrium. The FOSSA OVALIS is within the orifice of the postcava ; strictly speaking, what is here apparently an orifice of the postcava should be regarded as part of the atrium itself ; notwithstanding the description by Morrell and the observations embodied in several theses for graduation at Cornell University there are many points of comparison between the hearts of man and sheep that have not been made satisfactorily.

1, ridge at the junction of the PRECAVA.   2, transection of a prominent TRABECULA of the APPENDICAL PART of the AURICLE.   3, ectal surface of APPENDIX.   4, termination of cut of wall.   5, trabecula.   6, smooth ental surface of ATRIUM.   7, valve between POSTCAVA and CORONARY SINUS.   9, TUBERCLE OF LOWER between postcava and precava.

HEART OF THE SHEEP, THE LEFT SIDE OPENED ; × .8.

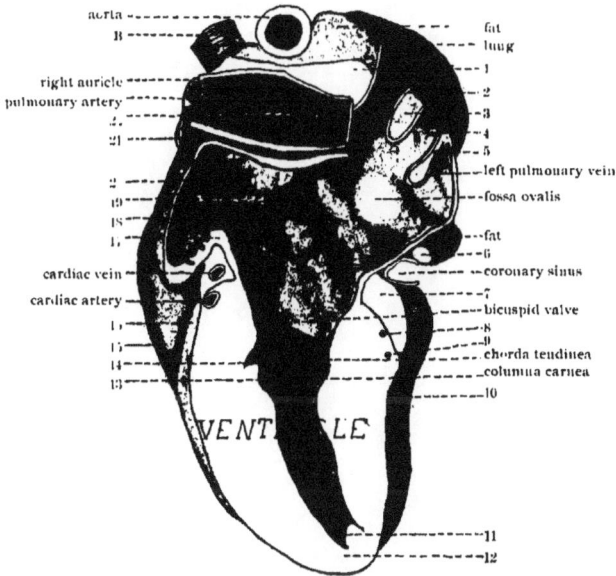

From this preparation (2789) were removed the left walls of the left auricle and ventricle, and of the pulmonary artery and coronary sinus. Most of the shading is too heavy.

1, cut surface of FAT. 2, 4, mouths of RIGHT PULMONARY VEINS. 3, PARTITION
between them. 5, partition between the right and left pulmonary veins ; it is made too
narrow, while the cut edge of the farther wall is too thick. 6, 7, cut fat at AURICULO-
VENTRICULAR SULCUS and about CORONARY SINUS. 8, 9, branches of CARDIAC
(coronary) ARTERY. 10, ectal surface of LEFT VENTRICLE. 11, apex of cavity of ventricle. 12, thin apical part of wall. 13, depression between the MUSCULAR RIDGES.
14, notch indicating the existence of a considerable space behind (ventrad of) that part of
the wall. 15, fat at base of ventricle. 16, muscular wall of ventricle. 17, ridge at junction of AURICLE and ventricle ; 18, 20, depressions between TRABECULÆ in appendical
part of auricle. 19, ectal surface of appendix. 21, part of ectal surface of PULMONARY
ARTERY. 22, place of division of pulmonary artery into the left branch, here seen continued for about 1 cm., and the right branch ; the shading is so heavy as to obscure the partition between the two.

Aside from the general relations of parts the special objects of this figure are as follows :
(a), to exhibit the great thickness of most of the left ventricular wall. (b), to show an
AURICULO-VENTRICULAR VALVE ; the TRICUSPIDS on the right side have the same
general character; the margin is held at the corners by the TENDINOUS CORDS attached
to the FLESHY COLUMNS. (c) to indicate the location of the FOSSA OVALIS, the thin
area of the INTER-AURICULAR SEPTUM which was open in the fetus as the FORAMEN
OVALE. (d) to designate the point of attachment of the DUCTUS ARTERIOSUS, the
remnant of a free communication between the pulmonary artery and the AORTA in the
fetus ; it is a slight depression at the end of the line from the name *pulmonary
artery*.

THE SALIVARY GLANDS OF THE CAT. From *Anatomical Technology.*

The skin has been dissected from most of the face and the neck mainly to expose the two larger SALIVARY GLANDS. The PAROTID (*Glandula parotis*) is so named from its proximity to the ear ; most of it has probably been cut off with the ear in the specimen prepared as in Pl. II At its cephalic border are seen several ducts converging to form one, the PAROTID DUCT or DUCT OF STENO ; near the corner of the mouth the duct pierces the cheek and opens opposite the largest maxillary tooth ; see Plates XVI and XVII. In man this duct opens opposite the second molar tooth, the last but one.

The SUBMAXILLARY GLAND (*Gl. submaxillaris*) lies caudo-ventrad of the parotid and is separated from it by the JUGULAR VEIN (*V. jugularis externa*) which forms a loop about it. Its DUCT OF WHARTON opens into the floor of the mouth (Fig. 15, *Ductus Wharton*). Just cephalad of the submaxillary are some LYMPHATIC GLANDS (*Gl. lym.*)

Only the parts here enumerated need be considered in this connection.

MANDIBLE, the lower jaw. *M. tmp.*, the temporal muscle. *M. mstr.*, the MASSE-TER MUSCLE. ZYGOMA, the zygomatic arch or cheek bone ; see Pl. XVI.

HEAD AND NECK OF CAT PARTLY DISSECTED

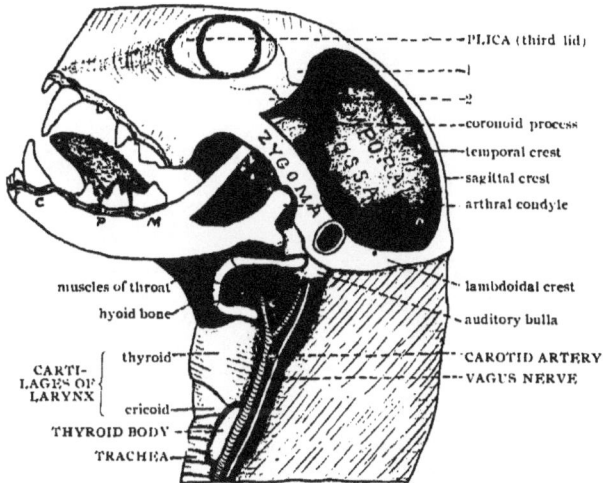

Compare with Plates I, and II. The tip of the nose has been cut off, and the muscles (TEMPORAL and MASSETER) removed ; they arise from the side of the CRANIUM and the ZYGOMA and are inserted upon the MANDIBLE (lower jaw) so as to close it vigorously. Some muscles have also been removed from the throat and neck so as to expose the LARYNX, the HYOID BONE, the TRACHEA, the CAROTID ARTERY, VAGUS (pneumogastric) NERVE and THYROID BODY ; in man the lateral lobes of the thyroid are connected across the trachea by an isthmus.

The nerve here shown really includes within one sheath two nerves, the vagus and the CERVICAL SYMPATHIC (sympathetic) ; for the sake of clearness they are not distinguished ; neither are their ganglia or branches shown ; only one branch of the carotid is indicated.

The TEMPORAL CREST is too near the meson in its cephalic part, so that the TEMPORAL FOSSA is made too extensive.

1, POSTORBITAL PROCESS of the frontal bone which is connected by ligament with the smaller projection (2) of the malar bone and thus incloses the ORBIT ; see Pl. I.

The tongue and papillæ are better shown in Pl. XVII.

The capital letters on the mandible indicate the four groups of TEETH.

C, the tusk-like CANINE, longer than the rest. I, the three INCISORS. P the two PREMOLARS (bicuspids). M, the single MOLAR.

In the maxilla (upper jaw) the canine is easily recognized ; only one incisor appears, the others being hidden behind it. The three other teeth are premolars. There is a single small molar hidden mesad of the last premolar.

MESAL ASPECT OF THE RIGHT HALF OF
THE CAT'S HEAD, SLIGHTLY ENLARGED. From
the *Reference Handbook*. Fig. 5087, somewhat modified ; the original figure, on a smaller scale, is in *Anatomical Technology*.

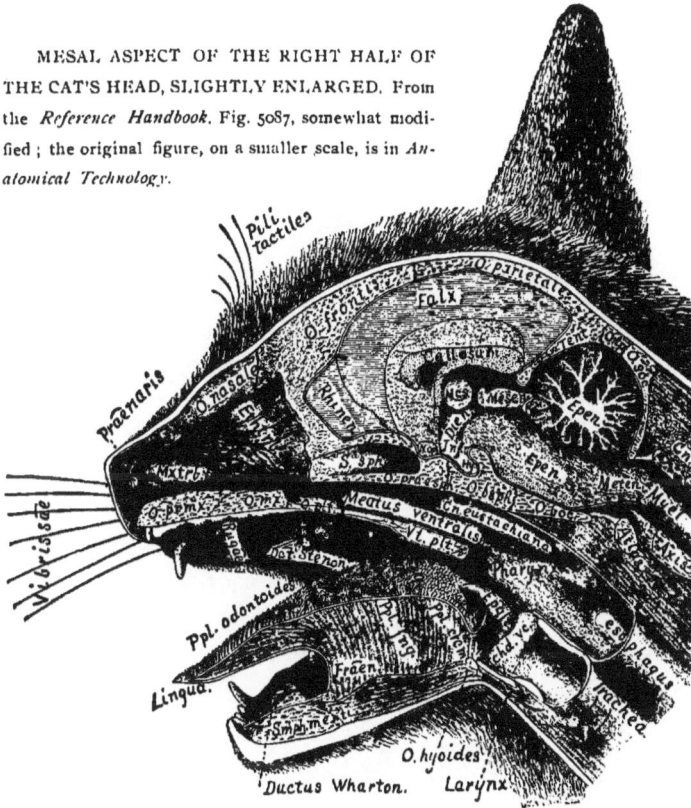

Most of the mesal parts are medisected, but in order to expose the right nasal cavity
the NASAL SEPTUM has been mostly cut away ; it may be seen in some figures of the
human head.

Cn., the neural, spinal or vertebral, canal ; it is represented by the dark areas dorsad
and ventrad of the myel, and lines thereto should have been drawn from the name.

Cn. (*Canalis*) *Eustachiana*, the orifice of the Eustachian tube ; it is represented by the
crescentic line just dorsad of the letters *Cn.*

Chd. vc., vocal cord or band. *Dct. Stenon*, duct of the parotid gland, indicated by the
bristle extending cephalad from the large tooth. *Ductus Wharton*, the duct of the submaxillary gland, opening at papilla. *Epgll.*, the epiglottis. *Lingua*, the tongue.
Meatus Ventralis, the ventral and more direct passage from the nostrils (prenares) to the
pharynx. *Myel*, the spinal cord. *Ppl. odontoides*, the odontoid or tooth-like papillæ
of the tongue. *Ppl. fng.*, the fungiform papillæ. *Ppl. crcm.*, the circumvallate papillæ.
Rugæ, the transverse ridges on the roof of the mouth. *Smph. menti*, the articulation at
the chin between the two halves of the mandible ; in man the union early becomes firm
bone, but in most cats the separation may be effected by cutting or pulling. *Vl. pll*, the
soft palate.

The other names and abbreviations may be disregarded.

THE BRAIN OF THE SHEEP, THE CEREBRUM SLICED TO NEAR THE LEVEL
OF THE CALLOSUM ; × 1.5.

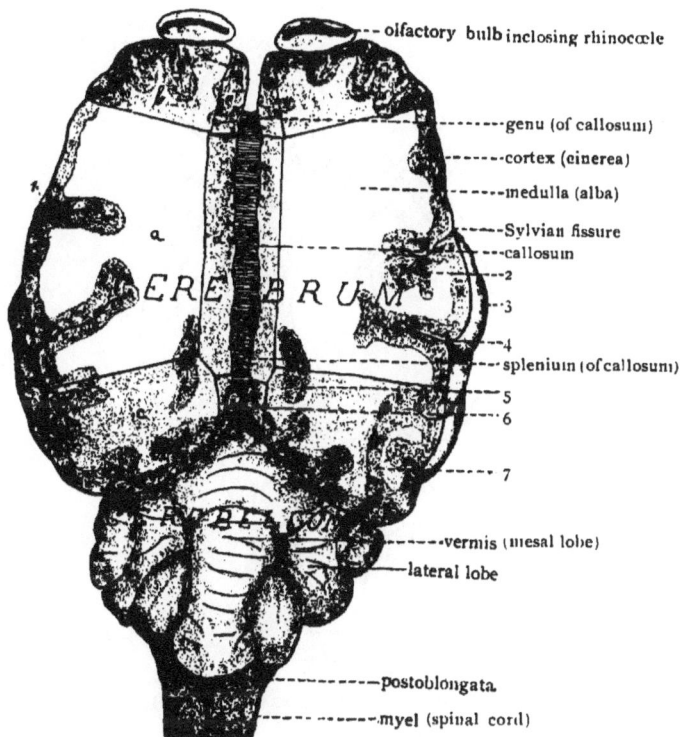

The following points are illustrated :

A. The general proportions of the two great masses, CEREBRUM and CEREBEL-
LUM.

B. The constitution of the cerebellum by a mesal lobe (VERMIS) and a pair of LAT-
ERAL LOBES.

C. The junction of the two halves of the cerebrum by a thick sheet of fibers, the
CALLOSUM ; its rounded cephalic and caudal margins are the GENU and SPLENIUM,
respectively ; Pl. XXIII.

D. The relative positions of the two kinds of substances in the larger part of the
cerebrum ; the ALBA (white substance) is central ; the CINEREA (gray substance) is peripheral, constituting the CORTEX.

E. The relation of the cortex to the FISSURES.

F. The passage of the ARACHNOID membrane across the mouths of the fissures, as
at 1 and 4, while the PIA dips to the bottom as a fold.

G. The existence of a cavity (RHINOCŒLE or olfactory ventricle) in the OLFAC-
TORY BULB ; see Pl. XXIV and p. 69, Fig. 3.

Defects.—The cerebellar divisions (FOLIUMS) are not shown in detail. In the dark
interval (6) between the cerebellum and cerebrum should appear the cut ends of vessels
one of which is shown in Pl. XXIV. There is no indication of the thin layer of cinerea on
the dorsum of the callosum ; Pl. XXV.

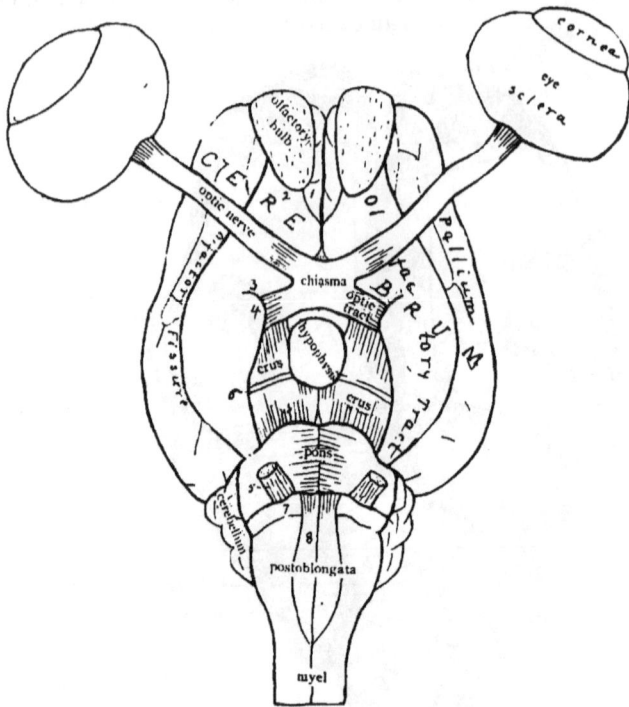

VENTRAL ASPECT OF THE SHEEP'S BRAIN WITH THE EYES AT-
TACHED; × 1.3.

From the specimens commonly examined the brain here represented differs as follows :
a. The EYES have been retained with OPTIC NERVES. *b.* The HYPOPHYSIS is
retained. *c.* Besides the TRIGEMINUS NERVES (marked 5), on the actual brains
there are more or less distinct signs of the roots of the other cranial nerves.

The numbers indicate parts as follows : 1, a small portion of the PALLIUM or fissured
region of the CERREBUM, projecting mesad of the OLFACTORY BULB. 2, the OL-
FACTORY CRUS, connecting the OLFACTORY TRACT with the BULB. 3, a slightly
depressed area just cephalad of the OPTIC TRACT. 4, a part which distinctly projects
over the tract. 5, The root of the TRIGEMINUS, the great sensory nerve of the face. 6,
a slight ridge, not always distinct, crossing the CRUS. 7, the TRAPEZIUM, concealed in
the human brain, by the caudal margin of the broad PONS. 8, the PYRAMID, less distinct than in man and not exhibiting a DECUSSATION.

Some details are more fully shown in Pl. XX.

BASE OF SHEEP'S BRAIN AFTER THE REMOVAL OF THE HYPOPHYSIS
AND PARTS OF THE CEREBRUM AND CEREBELLUM ; enlarged.

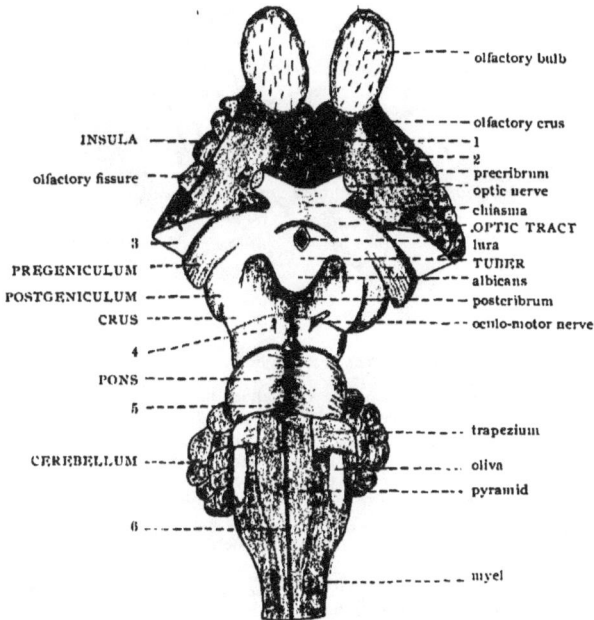

The cephalic and caudal regions are nearly the same as in Pl. XIX, but the following differences should be noted : *a*. The absence of the frontal parts of the cerebrum between and laterad of the OLFACTORY BULBS. *b*. The indication of the MESAL (1) and LATERAL (2) ROOTS of the bulb. *c*. Between the two the irregular triangular area, PRECRIBRUM ("anterior perforated space") presenting orifices for the transmission of vessels. *d*. The removal of the HYPOPHYSIS ; this exposes a slight elevation, TUBER, and an orifice, lura, leading into the diacœle. *e*. The CRURA and OPTIC TRACTS are more fully seen. *f*. The PONS presents more distinctly the mesal emargination of its caudal margin.

1, Mesal root of olfactory bulb. 2, lateral root. 3, cut surface of olfactory tract and pallium. 4, depression caused by the extraction of the right oculo-motor nerve. 5, Caudal emargination of the pons. 6, Ventral mesal sulcus of the oblongata. The unnamed shaded line across the crus just cephalad of the oculo-motor nerve was intended to represent the cimbia.

LEFT SIDE OF THE SHEEP'S BRAIN AFTER THE REMOVAL OF MOST OF
THE CEREBRUM AND CEREBELLUM ; × 1.

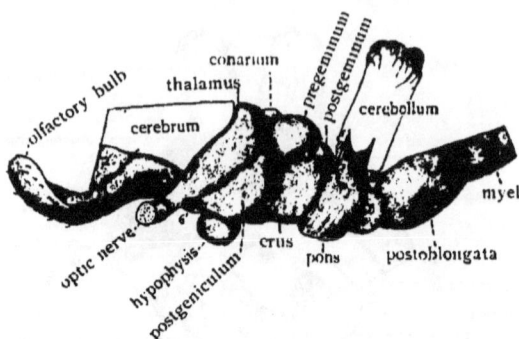

The CEREBELLUM is left of its natural height, but the cephalic and caudal convexities are sliced away so as to expose the parts which are overhung by them. In a companion preparation the dorsal portion of the cerebellum has also been removed, with the cephalic and caudal convexities, but the lateral "overhangs" are retained.

The CEREBRUM has been cut down to the level of the THALAMI ; the caudal portion cut away along the oblique line of its projection over the part marked 5 ; the lateral portion so as to expose the part marked 3 ; also the cephalic projection which, as seen in Plates XIX and XXV, overhangs the OLFACTORY BULBS.

The short lines on the surface of the olfactory bulb represent the OLFACTORY NERVES. The cut end of the left OPTIC NERVE is dotted to indicate its fibrous structure.

1, OLFACTORY CRUS ; compare with Pl. XX. 2, a part of the PALLIUM which has not been cut. 3, OLFACTORY TRACT. 4, (indistinct), CHIASMA. 5, PREGENICULUM (external or anterior geniculate body), distinct in man but here little more than a lateral portion of the thalamus. 6, TUBER (*cinereum*), the slight convexity to which the HYPOPHYSIS is attached ; in Pl. XX it is the area just caudad of the chiasma. 7, the MEDIPEDUNCLE, continuing the PONS to the lateral mass of the cerebellum. 6, the TRAPEZIUM ; compare with Pl. XX.

Excepting the unshaded areas, representing cut surfaces, all the parts seen in this figure were covered by PIA.

At the dorsal end of the cerebellum are seen a few FOLIA, its leaflet-like divisions ; these are not shown in any other plate.

Besides facilitating the recognition of certain important parts this figure well illustrates the *segmental constitution* of the brain, which is obscured in the entire organ by the preponderance of the cerebrum and cerebellum. There is a series of more or less distinct masses demarcated by constrictions of greater or less depth. Admitting that there is still some doubt as to number and limits of the segments the following assignments may be accepted provisionally :

Olfactory bulbs and crura, } RHINENCEPHAL.

Cerebrum } PROSENCEPHAL (fore brain).

Thalami, conarium, hypophysis, } DIENCEPHAL (inter-brain).
  chiasma, and geniculums,

Geminums and crura } MESENCEPHAL (mid-brain).

Cerebellum, pons and preoblongata, } EPENCEPHAL (hind-brain).

Postoblongata } METENCEPHAL (after-brain).

DORSUM OF SHEEP'S BRAIN AFTER THE REMOVAL OF PARTS OF THE CEREBRUM AND CEREBELLUM.

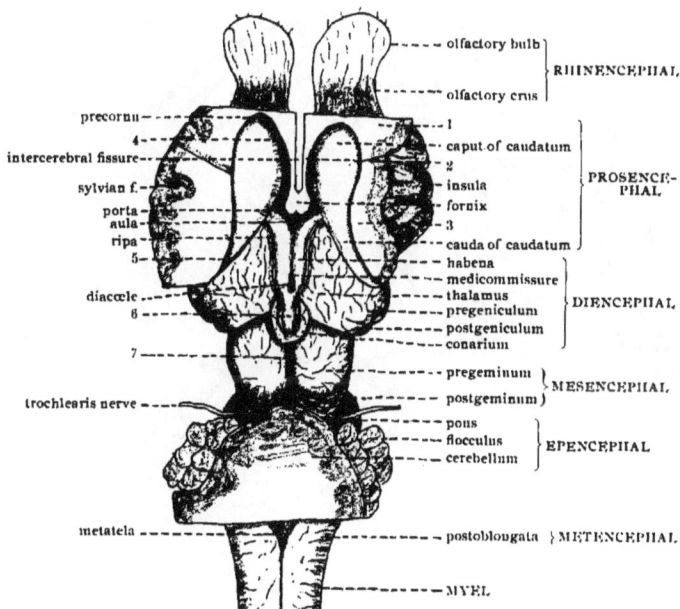

Compare with Pl. XVIII. From the cerebellum have been cut the dorsal part and also the caudal. On the cut dorsal surface are seen the central alba and the peripheral cinerea, but the outline of the latter is diagrammatic only. At the sides are the tiers of foliums constituting the flocculus.

From the cephalic end of the cerebrum have been cut the parts projecting over the olfactory crura, but part of the cephalic slope marked *b* in Pl. XVIII is here marked 1. With the dorsal portion were removed the entire callosum and the fornix excepting the cephalic vertical part. This and the mesal walls of the paracœle are really cut at a lower level than the larger cut surface on the left. On the right the insula has been exposed by pushing up and breaking off the overhanging parts. The ectal surfaces, covered by pia, are indicated by irregular lines representing the blood-vessels.

The ental surfaces, covered by endyma, are those of the caudatums in the paracœles, the habenas, medicommissure and conarial pouch ; and the floor of the aula and p rtas.

The irregular line laterad of the habena and extending around the endymal area on the conarium represents a ripa (shore-line). It consists of the cut or torn edges of the pia from the dorsum of the thalamus and of the endyma from the habena which united to form a membranous roof of the diacœle, the DIATELA, which has been removed.

Similarly the pial, dorsal surface of the thalamus is demarcated from the endymal surface of the caudatum by a ripa which meets the other at the porta.

The CONARIUM, although a constituent of the DIENCEPHAL, is tilted caudad so as to rest upon the PREGEMINUM, and the mesal part of its exposed surface is likewise covered by endyma.

1, cephalic slope. 2, mesal wall of PARACŒLE ; at a higher level this would be one of the HEMISEPTUMS. 3, horizontal cut surface of cerebrum. 4, the mesal, vertical portion of the paracœle. 5, indicates the location of the ripa between the thalamus and caudatum, but it is overhung by the latter so as not to appear in this view. 6, extension of the diacœle upon the conarium. 7, mesal furrow of the pregeminum.

MESAL ASPECT OF RIGHT HALF OF SHEEP'S BRAIN; × 2.

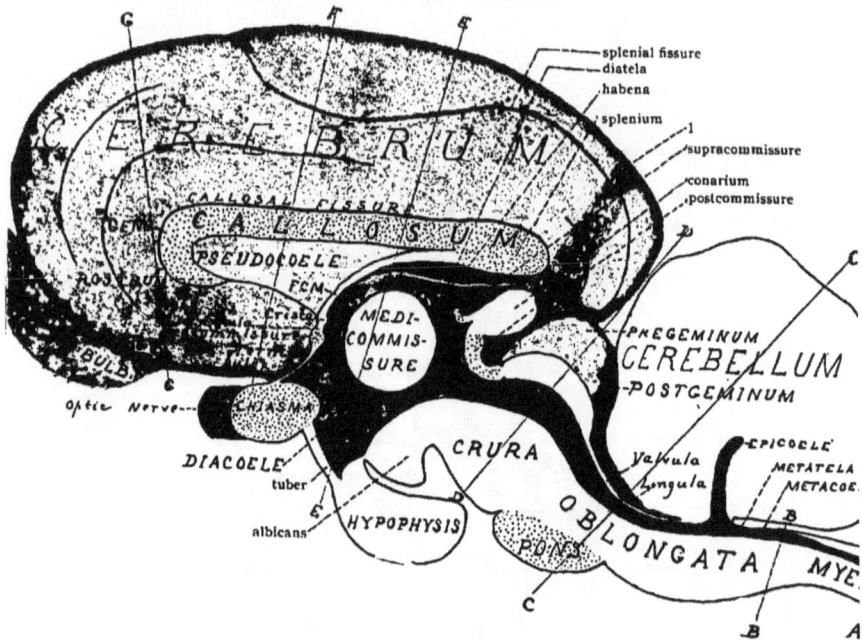

This figure is semi-schematic, certain details being omitted for the sake of clearness, *e. g.*, the divisons of the CEREBELLUM, the VESSELS, and the MEMBRANES, ARACH-NOID and PIA. The pia, however, is represented by the line between the ROSTRUM and the CRISTA.

The tuber is the *Tuber cinereum*, called *torus* in the former edition.

The objects of the figure are : To show most of the MESAL PARTS ; to illustrate ENDYMAL CONTINUITY and its concomitant, CŒLIAN CIRCUMSCRIPTION ; to indicate the PLANES OF TRANSECTION which are most instructive, *A—G*. Compare Plates XX, XXII, XXIV XXV. For fuller description see Practicum IX.

SHEEP'S BRAIN, THE PARACŒLES (lateral ventricles) EXPOSED ; × 2.

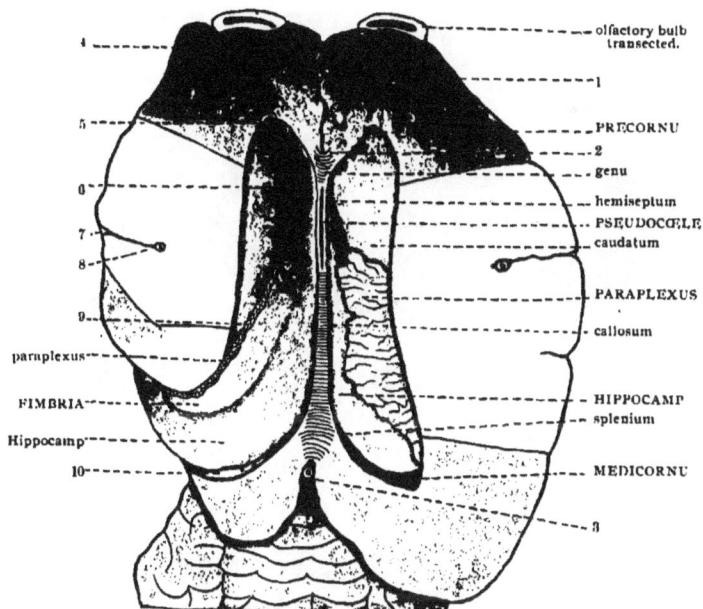

This figure represents a stage of dissection intermediate between Plates XVIII and XXII. By the removal of successive slices the PARACŒLES have been opened ; the left has then been more completely exposed by oblique sections, and the PARAPLEXUS trimmed off so as to expose the wide FIMBRIA and the furrow between it and the HIPPOCAMP. The plane of section did not coincide exactly with the CALLOSUM ; the caudal three-fifths of this is represented by the transverse lines ; also the cephalic end, the GENU ; but an intermediate portion is wholly removed, exposing the narrow PSEUDOCŒLE ("fifth ventricle") and its thin lateral walls, HEMISEPTUMS. The HEMISEPTUM is here shown to be only a portion of the general mesal wall of the paracœle. The Pseudocœle (Pl. 25) has no connection with the true cavities of the brain. The only communications of the paracœles are through the PORTAS with the mesal AULA (Pl. XXII).

The two FIMBRIAS and HIPPOCAMPS connected by a mesal part (Pl. XXV) constitute the FORNIX.

The HIPPOCAMP is sometimes called *hippocampus major*.

1, INTERCEREBRAL FISSURE. 2, CALLOSAL FISSURE. 3, VESSEL. 4, interrupted lines indicating the continuation of the paracœle into the RHINOCŒLE. 5, PRECORNU. 6, CAPUT of the CAUDATUM. 7, SYLVIAN FISSURE crossed by ARACHNOID. 8, VESSEL at bottom of fissure. 9, CAUDA of CAUDATUM. 10, part of caudal wall of paracœle.

A. TRANSECTION OF SHEEP'S BRAIN ; × 1.5.

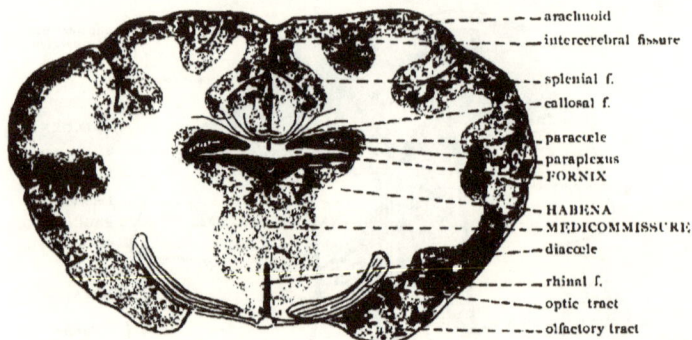

arachnoid
intercerebral fissure

splenial f.
callosal f.

paracœle
paraplexus
FORNIX

HABENA
MEDICOMMISSURE
diacœle

rhinal f.
optic tract
olfactory tract

The plane of section approximates *E* in Pl. XXIII. The HYPOPHYSIS has been removed and the DIACŒLE is open ventrad at the LURA (Pl. XX). The OPTIC TRACT is cut obliquely ; its fibrous structure is roughly indicated by lines. The masses connected by the MEDICOMMISSURE are the THALAMI. The CALLOSUM is indicated by lines. The INTERCEREBRAL FISSURE is bridged by the ARACHNOID ; in man the falx descends into the fissure for a greater or less distance carrying the arachnoid before it. The fissure here named *rhinal* is named *olfactory* in Pl. XIX.

B. ENLARGEMENT OF THE CENTRAL REGION OF A.

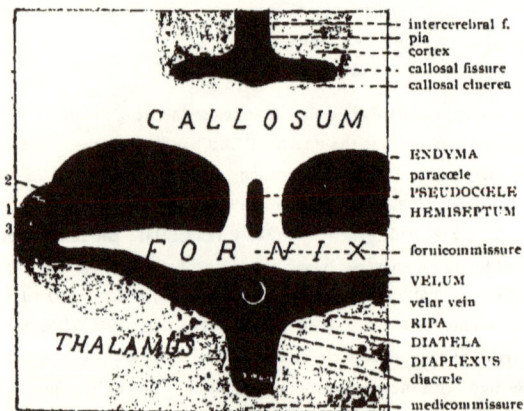

intercerebral f.
pia
cortex
callosal fissure
callosal cinerea

CALLOSUM

ENDYMA
paracœle
PSEUDOCŒLE
HEMISEPTUM

FOR-N-I-X
fornicommissure

VELUM
velar vein
RIPA
DIATELA
DIAPLEXUS
diacœle

THALAMUS

medicommissure

This was designed to exhibit more clearly the relations of the CAVITIES to the MEMBRANES and PLEXUSES, but some points are obscured by the shading
The mesal DIACŒLE and the lateral PARACŒLES are lined by smooth ENDYMA, represented by a heavy line. In the diacœle the endyma may be traced dorsad upon the mesal surface of the THALAMUS and over the dorso-mesal ridge representing the HABENA, as far as the point called RIPA, (see Pl. XXII, left side.) Here it is reflected mesad upon the ventral surface of the VELUM
The velum consists of the PIA covering the ventral surface of the FORNIX and the dorsal surface of the thalami, together with CONNECTIVE TISSUE and VESSELS (of which only one is shown). Near the meson there hangs into the diacœle at each side a plexus (DIAPLEXUS) covered by the endyma.
At the interval (RIMA) between the margin of the fornix and the caudatum (1) the velum extends into the paracœle as the PARAPLEXUS, covered, however, by the endyma which is reflected off at 3 and the point opposite.
A thin layer of the cortex extends across the callosum.